APPLIED FLUID FLOW MEASUREMENT

ENGINEERING MEASUREMENTS AND INSTRUMENTATION

A Series of Reference Books and Textbooks

Editor

Paul N. Cheremisinoff

New Jersey Institute of Technology

Newark, New Jersey

Volume 1. Applied Fluid Flow Measurement: Fundamentals and Technology, *by Nicholas P. Cheremisinoff*

Additional Volumes in Preparation

APPLIED FLUID FLOW MEASUREMENT

Fundamentals and Technology

NICHOLAS P. CHEREMISINOFF

MARCEL DEKKER, INC. New York and Basel

Library of Congress Cataloging in Publication Data

Cheremisinoff, Nicholas P.
Applied fluid flow measurement.

(Engineering measurements and instrumentation; v. 1)
Includes bibliographical references and index.
1. Fluid dynamic measurements. 2. Flow meters.
I. Title. II. Series.
TC177.C42 532'.051'028 79-19138
ISBN 0-8247-6871-X

MARCEL DEKKER, INC.
270 Madison Avenue, New York, New York 10016

Current printing (last digit):
10 9 8 7 6 5 4 3 2

PRINTED IN THE UNITED STATES OF AMERICA

Preface

Throughout history man has struggled for an understanding of the flow of fluids and strived to invent methods of accurate flow measurement in order to harness and control its benefits. The fundamental principles of fluid flow behavior, postulated centuries ago by men such as Benedetto Castelli, Evangelista Torricelli, John Bernoulli, Henri Pitot, Giovanni Battista Venturi, and many others, form the basis of our current understanding of fluid flow and many of our modern-day measuring techniques.

Today, research and industry rely heavily on accurate flow measurement to supply data for analysis and systems performance assessment. The accuracy required of flow measurement can range from crude estimates to extreme precision. The proper selection of a flow measuring instrument can depend on numerous factors, including specific application and cost.

In many applications, accurate measurement is directly related to profits. Small errors in flow measurement can be the cause of many thousands of dollars lost over a period of time.

Accurate flow measurement has a direct impact on energy conservation as well. In the automotive industry, for example, very precise flow measurement is desirable to achieve reductions in emissions and fuel consumption.

In the area of environmental control, the handling and treating of effluents is one of industry's major and, in many cases, most expensive problems. With laws and federal regulations becoming stricter, both industries and municipalities are finding it necessary to accurately measure and maintain records of their discharges. As such, flow measurement has become an essential operation in the control and treatment of wastestreams.

This book provides an overview of fluid flow measuring devices employed throughout industry. A detailed description of the principles of operation of the various flow measuring instruments is provided. A working knowledge of the operational principles behind each device is essential if the proper instrument selection is to be made for a particular application. The book is intended as a working reference for practicing engineers and as a supplemental text for engineering students.

My heartfelt thanks go to many friends and those in industry who gave of their valuable time, expertise, and materials, making this volume possible.

Nicholas P. Cheremisinoff

Contents

1

Review of Fluid Behavior and Characteristics

1.1 INTRODUCTION

Accurate measurement of gas and liquid flow is important for a variety of processes and reasons. Accurate measurement is needed in obtaining specific proportions. It is important in maintaining specific rates of flow, without which precise quality control would not be possible.

If erroneous flow measurements are made, cost estimates based on flow data will be incorrect. In many processes, huge volumes of gas, steam, and liquids are often measured daily; as such, small percentage errors in measurements can total to large sums.

In the area of pollution control, accurate flow measurement is essential to the analysis of a process's or plant operation's environmental impact. Accuracy in flow measurement is essential when collecting wastewater samples from streams, manholes, or other sources at precise timed or flow-proportioned intervals. Its importance cannot be overemphasized in measuring effluent rates from stacks.

There are a variety of schemes and commercially available devices for measuring flow quantity. The selection of a specific device should be based on: (a) the degree of accuracy and/or precision required for a particular application; (b) the particular scheme best suited to the process application and conditions; and (c) a comparison of the limitations and costs of various commercial units.

The purpose of this book is to provide an overview of the alternatives available to flow measurement. Fundamentals behind the operation of various measuring principles are stressed and manufacturer's data and recommended applications included wherever possible.

Before an understanding of the operational principles of various flow measuring devices can be gained, the basics of fluid flow should be reviewed. A fluid is any material which undergoes continuous deformation upon being subjected to a shearing stress or force. The resistance to deformation or flow is the consistency or viscosity of the fluid. A perfect or ideal fluid is a liquid or gas that has no resistance to shear. For gases and Newtonian liquids, viscosity is constant if the static pressure and temperature are fixed. Unfortunately, there are a large number of fluids that do not follow the behavior of ideal gases and simple liquids (Newtonians). Many slurries encountered in wastewater treatment, for example, approximate the behavior of plastic-viscous materials. An understanding of the rheological properties and flow behavior of various fluids is important to the selection of a flow measuring device.

1.2 PROPERTIES AND BEHAVIOR OF NEWTONIAN FLUIDS

Rheology can be defined as the science of deformation and flow of matter. It is helpful in explaining and measuring the flow conditions that are encountered for almost every type of natural or synthetic material.

The basic law of viscosity was deduced by Sir Isaac Newton, who described the flow of a fluid over a solid boundary surface as illustrated in Figure 1.1. In the figure, the fluid, either gas or liquid, is positioned

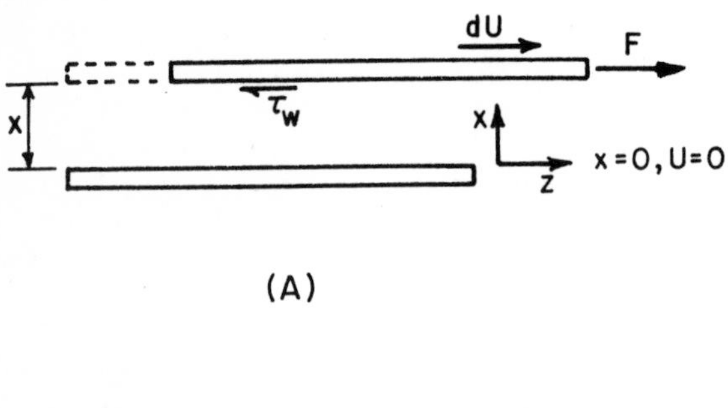

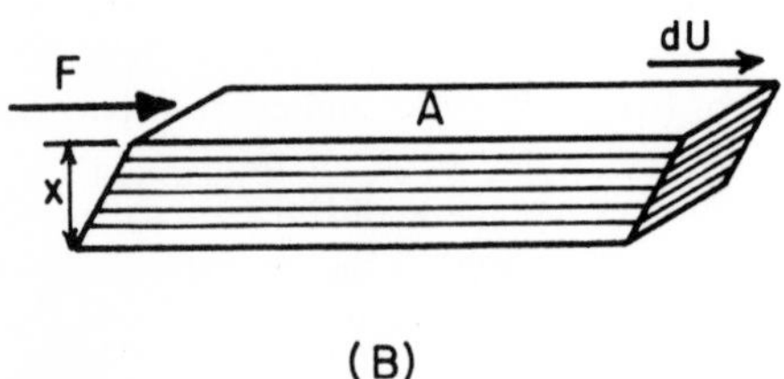

Figure 1.1 (A) Illustrates the system under consideration: laminar flow of Newtonian fluid. (B) The principle can be demonstrated by pressing downward on a deck of cards with a simultaneous forward push.

between two flat plates at distance x apart, and a force, F, is applied in the plane of the upper boundary so as to maintain a constant velocity du relative to the stationary lower plate. If force F is at steady state, the flow is laminar and F becomes balanced by a shearing force at the top solid-fluid interface. The shearing force results from the viscosity of the fluid. The fluid at the lower boundary is assumed to be stationary ("no slip" condition), with adjacent layers of fluid increasing in velocity as the distance from the stationary surface increases. (The principle can be physically described by the condition observed by pressing downward on a deck of playing cards with a simultaneous forward push. Figure 1.1B illustrates the force applied and the relative motion of the cards[1].)

The principle can be described mathematically by Eq. (1.1):

$$F = \mu A \frac{du}{dx} \qquad (1.1a)$$

where A is the area of the plate and du/dx is the velocity gradient or shear rate. The proportionality constant μ is the viscosity of the fluid.

Shear stress is defined as the force per unit area:

$$\tau_{xz} = \frac{F}{A} \qquad (1.1b)$$

The first subscript on τ (the x direction) denotes the orientation of the surface on which the stress acts, and the second subscript (the z direction) indicates the direction in which the stress applies. (That is, τ_{xz} is the shear stress that acts in the z direction on a surface normal to the x direction.)

Fluids that exhibit a rate of flow or shear that is directly proportional to the stress or shear force are termed Newtonian fluids (refer to Figure 1.2). Rearranging Eq. (1.1), Newton's law of viscosity is obtained, which describes the direct proportionality between shear stress and shear rate of fluids in laminar flow:

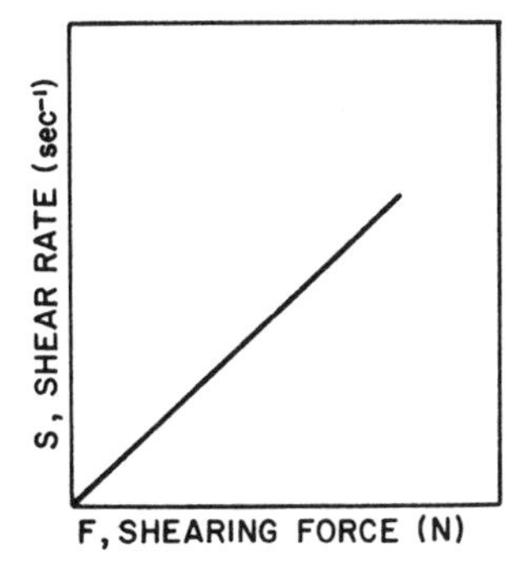

Figure 1.2 Illustrates the characteristics of a Newtonian fluid.

$$\tau_{xz} = \frac{\mu}{g_c} \frac{du}{dx} \tag{1.2}$$

If the axis of orientation is from the top plate to the bottom in Figure 1.1A, Newton's law of viscosity becomes

$$\tau_{xz} = -\frac{\mu}{g_c} \frac{du}{dx} \tag{1.3}$$

Here, τ_{xz} can be interpreted as the viscous flux of z momentum in the x direction. With materials that are Newtonian, energy is dissipated by the collision of relatively small molecules.[2,3]

For incompressible fluids (most liquids), the effect of variations of system pressure and temperatures on density is often considered negligible under actual flow conditions.

For an incompressible fluid, the expression becomes

$$\tau_{xy} g_c = -\frac{\mu}{\rho} \frac{d(u\rho)}{dx} \tag{1.4}$$

where the quantity $u\rho$ has the dimensions of momentum per unit volume (often called momentum concentration).

The viscosity of a Newtonian fluid is independent of the rate of shear. Figure 1.3 illustrates the relationship of viscosity to changing shear rate. For a material classified as Newtonian, then, a one-point measurement is sufficient to obtain the shear rate versus shearing force relationship (Figure 1.2).

All Newtonian liquids will experience a decrease in viscosity with increasing temperature at constant pressure. For gases, viscosity will increase with increasing temperature at a constant pressure. (This follows

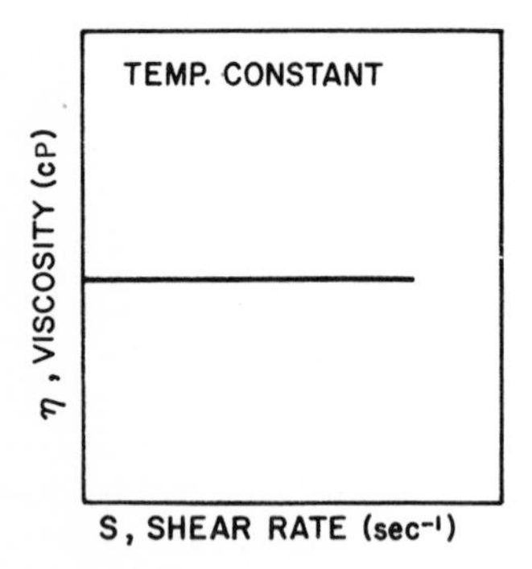

Figure 1.3 Illustrates that the viscosity of a Newtonian fluid is independent of shear rate.

from the kinetic theory of gases which postulates viscosity to be proportional to density, the molecules' average velocity, and the mean free path of molecules.)

For many liquids, viscosity increases with pressure at a constant temperature. Below the critical pressure, however, the pressure dependency is small. For gases, viscosity also increases with increasing pressure. This is in contradiction to the kinetic theory which predicts viscosity to be independent of pressure (for real gases at high reduced temperatures and low reduced pressures, the theory is applicable). Viscosity data for fluids have been correlated by Uyehara and Watson[4] on a reduced basis (refer to Figure 1.4). The curves shown in Figure 1.4 are a function of the reduced pressure P/P_c. The critical viscosity can be estimated from Eq. (1.5), proposed by Uyehara and Watson[4]:

$$\mu_c = \frac{61.6\sqrt{MT_c}}{V_c^{2/3}} \tag{1.5}$$

where M is the molecular weight, T_c the critical temperature (K), V_c the critical volume ($cm^3/g{\cdot}mol$), and μ_c the critical viscosity (centipoise, cP).

1.3 PROPERTIES AND BEHAVIOR OF NON-NEWTONIAN FLUIDS

Non-Newtonian fluids are materials that show nonlinearity in the shear rate versus shearing force relationship (refer to Figure 1.5). For non-Newtonians, the term "apparent viscosity" is used to describe a material's resistance to flow. Apparent viscosity is defined simply as the viscosity the material would have if it were a Newtonian fluid.

Non-Newtonians are divided into three categories that, depending on the material, often overlap

1. Time-independent fluids
2. Time-dependent fluids
3. Viscoelastic

Time-Independent Fluids

These are fluids in which the shear rate at a given point is dependent only on the instantaneous shear stress resulting. Time-independent non-Newtonians are often called non-Newtonian viscous fluids.

The term "plastic flow" refers to materials that exhibit a yield value. Yield value is the shear stress or force that must be applied to a material to induce flow. The flow curves of various types of time-independent materials are shown in Figure 1.6. (Specific examples of materials exhibiting a yield value are ketchup, mustard, toothpaste, paper pulp, and mayonnaise.)

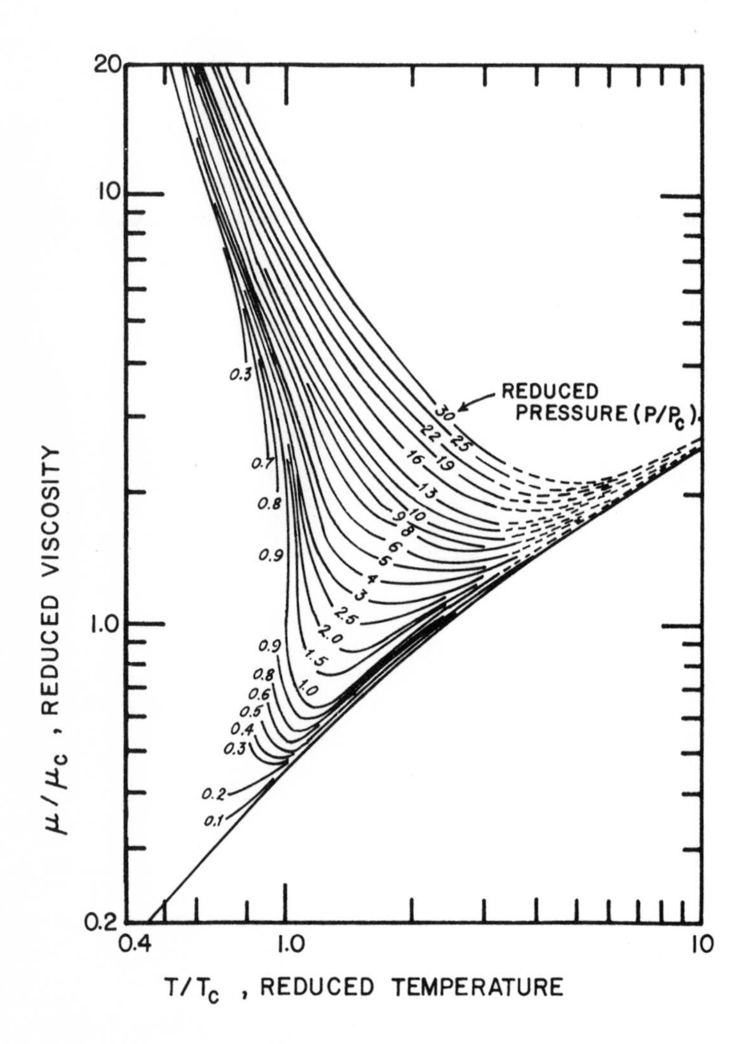

Figure 1.4 Generalized curve for estimating reduced viscosities of Newtonian fluids. (From Uyehara and Watson.[4])

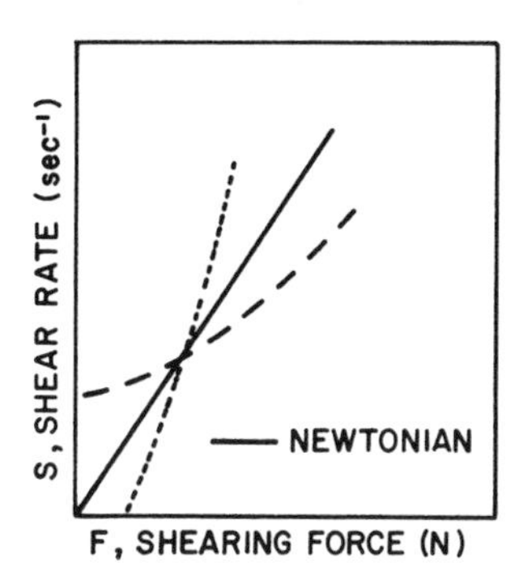

Figure 1.5 Comparison of the characteristics of a Newtonian fluid to non-Newtonian materials.

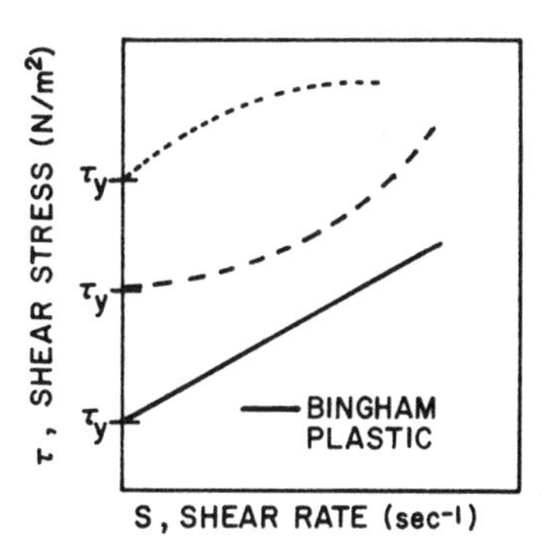

Figure 1.6 Flow curves of various classes of time-independent materials.

Analogous to the description of Newtonian fluids, apparent viscosity is defined as[3,5-9]

$$\mu_a = \frac{\tau_{xz} g_c}{du/dx} \tag{1.6}$$

According to Eq. (1.6), fluids with a yield stress will experience a decrease in apparent viscosity with increasing shear rate, as shown in Figure 1.7.

There are also a large group of non-Newtonian viscous fluids that do not exhibit a yield stress. Figure 1.8 shows various flow curves for these materials. Pseudoplastic fluids are characterized by linearity at very low and high shear rates in the curve shown in Figure 1.8. Pseudoplastic materials are often called power law fluids as shear stress follows the form of Eq. (1.7),

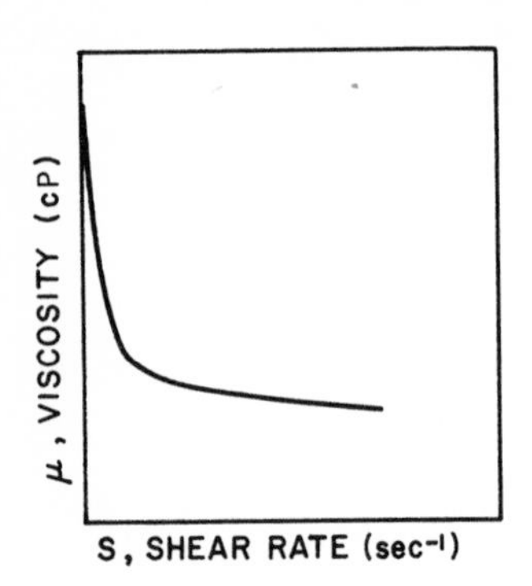

Figure 1.7 Viscosity decreases with increasing shear rate for time-independent fluids with a yield value.

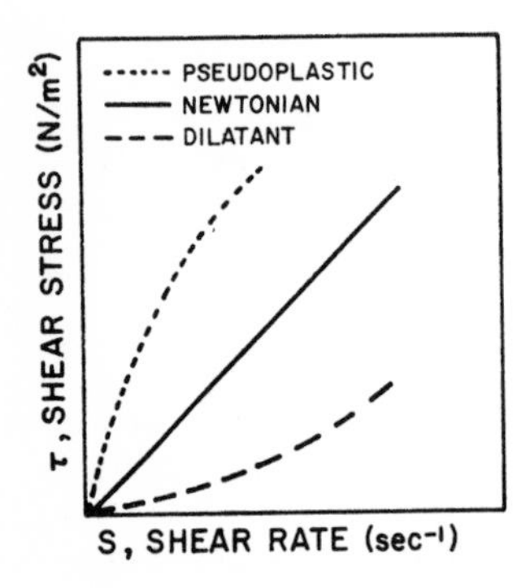

Figure 1.8 Flow curves for various types of fluids without a yield value.

$$\tau_{xz} = \frac{K}{g_c} \left| \frac{du}{dx} \right|^{n-1} \frac{du}{dx} \tag{1.7}$$

and apparent viscosity follows:

$$\mu_a = K \left| \frac{du}{dx} \right|^{n-1} \tag{1.8}$$

For these materials, the flow behavior index n is less than unity. Equation (1.8) indicates that apparent viscosity decreases with increasing shear rate. Examples of pseudoplastics include various polymer solutions, detergent slurries, paints, and greases.

Dilatant fluids are described as those which exhibit an isothermal increase in consistency on the apparent viscosity with increasing shear rate, as shown in Figure 1.9. These materials can often be characterized by a

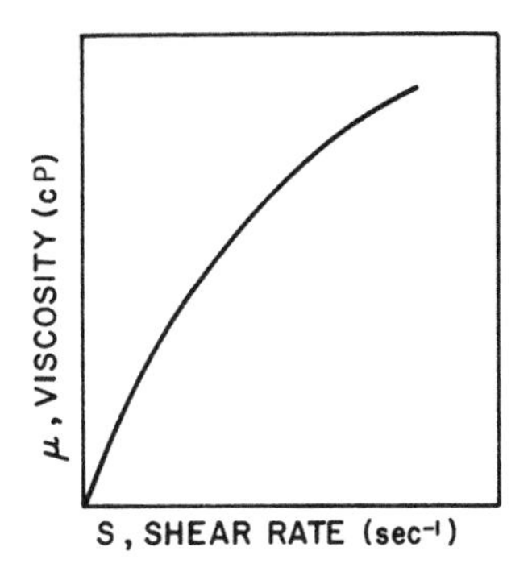

Figure 1.9 For dilatant materials, apparent viscosity increases with increasing rate of shear.

power law equation [Eq. (1.8)] for $n > 1$. There are two phenomena which these fluids exhibit: one is volumetric dilatancy in which the material increases in total volume with increasing shear rate; the other is the rheological dilatancy previously described.

Time-Dependent Fluids

The term "thixotropy" refers to an isothermal gel-sol-gel transformation or breakdown of a reversible colloidal gel. Figure 1.10 shows the characteristics flow curve for a thixotropic fluid. A curve such as this is obtained by first applying an increasing shearing force and then decreasing the force. The resultant curve shown is called a hysteresis loop, formed because of the finite time increment between the upward and downward portion of the curve. For a different time increment, the position of the loop will be different. Figure 1.11 shows the dependence of apparent viscosity with time at a constant shear rate. Green[10] has attempted to define an index of thixotropy based on the area encompassed by the loop.

Rheopectic fluids are a special form of thixotropic materials. The term "rheopexy" is often confused with dilatancy as viscosity appears to increase rapidly when the material is sheared slowly as compared to its stationary state.

In general, rheopectic fluids are rare. A typical hysteresis loop is shown in Figure 1.12. The concavity of the loop distinguishes these materials from the thixotropic fluids. The loop's location for a specific material will again depend on the fluid's time history, as well as the rate at which shear rate (du/dx) is increased and decreased.

Viscoelastic Materials

Viscoelastic fluids display properties typical of both viscous and elastic materials. These fluids will flow when subjected to stress; however, a portion

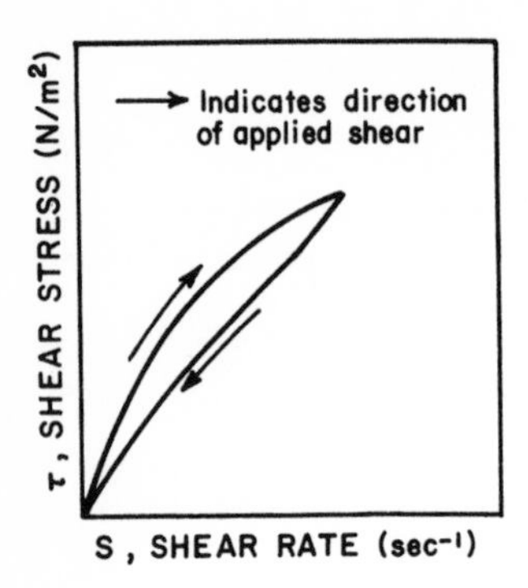

Figure 1.10 A hysteresis curve is formed for time-dependent non-Newtonian fluids. If the material is thixotropic, the upper portion of the curve is obtained by increasing the shear rate while the lower is generated by decreasing shear rate.

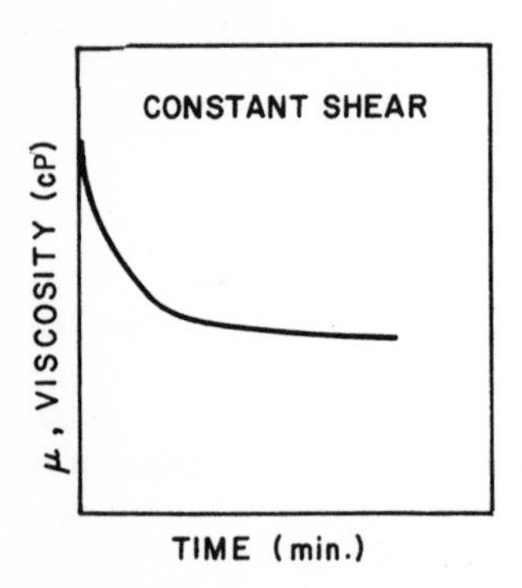

Figure 1.11 Viscosity decreases as a function of time under constant shear for thixotropic fluids.

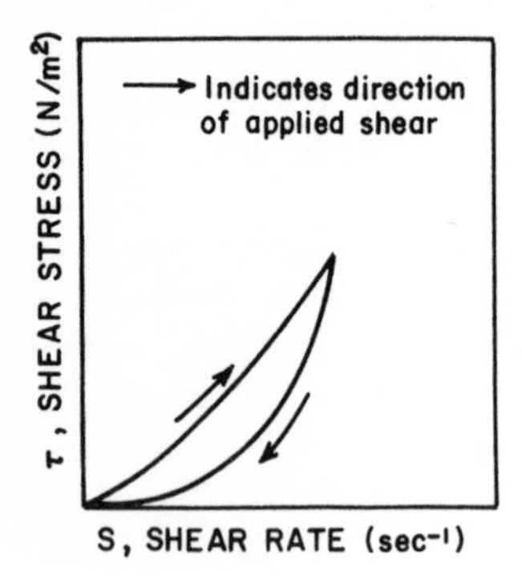

Figure 1.12 Hysteresis loop for a rheopectic fluid.

of their deformation gradually recovers upon removal of the stress.[2] The rheological properties of such materials depends on the relationships between shear stress, shear rate, and the time derivatives of these parameters.[2] Examples of these fluids are nylon, various jellies, and flour dough.

NOMENCLATURE

A	Area, ft^2 or m^2
F	Force, lb_f or newtons (N)
g_c	Conversion factor, 32.174 $lb_m \cdot ft/lb_f \cdot s^2$ or 4.17×10^8 lb_m $ft/lb_f \cdot h^2$
K	Fluid consistency index in power law, Eqs. (1.7) and (1.8), $lb_m \cdot s^{n-2} \cdot ft^{-1}$
M	Molecular weight
n	Flow behavior index in power law, Eqs. (1.7) and (1.8)
P	Pressure, lb_f/ft^2 or N/m^2
P_c	Critical pressure, lb_f/ft^2 or N/m^2
T	Temperature (°K)
T_c	Critical temperature (°K)
u	Local linear velocity in the z direction, ft/s or ft/h or m/s
V	Volume, cm^3
V_c	Critical volume, $cm^3/g \cdot mol$
x	Distance, ft or m
x, z	Coordinate directions

Greek Letters

μ	Viscosity, $lb_m/ft \cdot s$ or $lb_m/ft \cdot h$ or centipoise (cP)
μ_a	Apparent viscosity, ratio of shear stress to shear rate, $lb_m/ft \cdot s$ or $lb_m/ft \cdot h$ or cP
ρ	Density, lb_m/ft^3 or g/cm^3
τ	Shear stress, lb_f/ft^2 or N/m^2

QUESTIONS

1.1 What are the flow characteristics of a Newtonian fluid?

1.2 What is Newton's law of viscosity?

1.3 What are the three general categories of non-Newtonian fluids?

1.4 What does the term plastic flow mean?

1.5 Define apparent viscosity.

1.6 What is the general form of the equation that describes pseudoplastic fluids?

1.7 What does dilatancy mean?

1.8 What does thixotropic mean?

1.9 What is the range of values for the flow behavior index for pseudoplastic fluids?

1.10 What is the range of values for the flow behavior index for dilantant fluids?

1.11 What is the difference between a rheopectic fluid and a thixotropic fluid?

1.12 What is a hysteresis loop?

1.13 Name three viscoelastic fluids.

1.14 Give examples of pseudoplastic fluids.

1.15 Give examples of fluids exhibiting a yield value.

1.16 What is yield stress?

1.17 What two phenomena do dilatant fluids exhibit?

1.18 What is the "no-slip condition"?

1.19 What is the difference between rheopexy and dilatancy?

1.20 How does the viscosity of a Newtonian liquid vary with temperature at constant pressure?

1.21 How does the viscosity of a gas vary with temperature at constant pressure?

1.22 How does the viscosity of a liquid vary with pressure at a constant temperature?

1.23 How does the viscosity of a gas vary with pressure at a constant temperature

1.24 According to the kinetic theory of gases, what parameters is viscosity proportional to?

1.25 Define rheology.

2

Fundamentals of Fluid Flow

2.1 INTRODUCTION

In Chapter 1 viscosity, the physical property characterizing the flow resistance of fluids, was discussed. This chapter provides a review of fluid properties and the laws governing the motion of fluids.

The first part of this chapter deals with the fundamental conservation principles of mass and energy. Emphasis is on incompressible fluids (i.e., liquids), and various methods of evaluating the friction head loss are discussed. The latter portion of this chapter provides an overview of the models that attempt to describe the rheological and flow properties of non-Newtonian fluids.

It is assumed that the reader has a basic understanding of fluid mechanics, and so derivations are not stressed in the discussions that follow.

2.2 CONTINUITY AND THE LAWS OF CONSERVATION

The principle of conservation of mass states that mass can neither be created nor destroyed. If we consider the flow through the control volume shown in Figure 2.1 and apply this principle, the equation of continuity can be expressed as:

$$\rho_1 u_1 A_1 = \rho_2 u_2 A_2 \tag{2.1}$$

where

ρ = fluid density (gm/cm^3),

A = cross-sectional flow area (cm^2),

u = fluid velocity (cm/s)

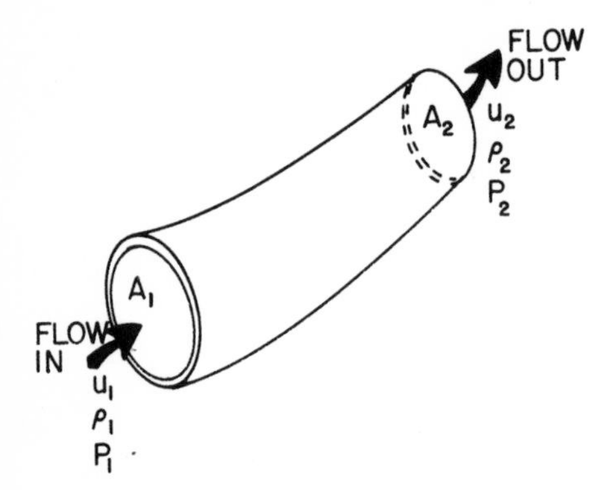

Figure 2.1 The system under consideration in this chapter.

Equation (2.1) describes steady flow which means that the discharge or rate of flow at any cross-section is constant. For an incompressible fluid, $\rho_1 = \rho_2$ and $A_1u_1 = A_2u_2$ [i.e., $Q_1 = Q_2$, where Q is the volumetric flow rate (cm^3/s)].

To determine the volumetric flow rate in a circular duct, where fluid velocity is a known function of the radial distance from the centerline axis (in the direction of flow), the following general equation is applicable.

$$Q = \int_0^R u(r)(2\pi r\, dr) \tag{2.2}$$

where R is the inside radius of the pipe (cm).

Energy Equation

Fluids in motion may possess four forms of energy:

1. Displacement or pressure energy
2. Velocity energy (kinetic energy)
3. Potential energy
4. Thermal or internal energy

The first three are of most interest in the subject of hydraulics.

In simple terms, energy can be defined as the ability to do work where work results from a force moving through a specified distance. In fluid flow, it is common to deal with energy per unit mass of fluid (N·m/kg in SI units or lb·ft/slug in English units).

The total of the first three types of energy can be expressed as follows:

$$E = E_p + E_k + E_p \tag{2.3}$$

where E_p is the energy per unit mass due to the elevation of the fluid (potential energy), defined as:

$$E_p = gz \tag{2.4}$$

Here g is the acceleration due to gravity (981.4 cm/s^2); z is a vertical distance above some datum (cm). A mass of fluid possesses potential energy that is equivalent to its ability to do work in falling through the distance z (work equals the fluid's mass times the acceleration due to gravity).

Note that E_k is the kinetic energy per unit mass of fluid defined as

$$E_k = \frac{1}{2}U^2 \tag{2.5}$$

and E_p is the energy per unit mass resulting from pressure applied to the fluid, defined as:

$$E_p = \frac{P}{\rho} \tag{2.6}$$

where P is pressure (N/m^2 or lb_f/ft^2).

Depending on the specific arrangement in a flow system (that is, if there is a pump, fan, turbine, or compressor in the flow system), mechanical energy can either be transferred to or from the fluid. In addition, whenever a fluid in motion passes a fixed boundary, friction exists. Fluid friction converts useful flow energy into heat or other forms of energy. The law of conservation of energy as applied to a fluid flowing between fixed points 1 and 2 can be stated as:

$$E_1 + E_m = E_2 + E_h \tag{2.7}$$

where $E_{1,2}$ is defined by Eq. (2.3); E_m represents the mechanical energies of the system; and E_h represents the energies lost per unit mass of moving fluids. Equation (2.7) can be more explicitly expressed as:

$$gz_1 + \frac{u_1^2}{2} + \frac{P_1}{\rho_1} + E_m = gz_2 + \frac{u_2^2}{2} + \frac{P_2}{\rho_2} + E_h \tag{2.8}$$

For an ideal or frictionless fluid, Eq. (2.8) reduces to the well-known Bernoulli equation:

$$z_1 + \frac{u_1^2}{2g} + \frac{P_1}{g\rho} + E'_m = z_2 + \frac{u_2^2}{2g} + \frac{P_2}{g\rho} + E'_h \tag{2.9}$$

where $E'_m = E_m/g$ and $E'_h = E_h/g$.

Each term in the above equation has the dimensions of energy per unit weight. The first term on the right-hand side of the expression represents the elevation head, the second is the velocity head, and the third is the pressure head.

Using general terms, piezometric (also referred to as hydraulic head) is defined as

$$H_P = z + \frac{P}{g\rho} \tag{2.10}$$

and total or stagnation head is defined as

$$H_T = z + \frac{P}{g\rho} + \frac{u^2}{2g} \tag{2.11}$$

Application of these definitions and the Bernoulli equation in pipeline systems is illustrated in Figure 2.2. The hydraulic grade line shown represents the piezometric head; the energy line represents the total head.

Momentum Equation

The law of conservation of momentum states that the net force applied to a mass of fluid is equal to its rate of change of momentum with respect to time. From Figure 2.1, this can be expressed as the following:

$$\Sigma F = \frac{(\rho_2 A_2 u_2 \, dt)u_2 - (\rho_1 A_1 u_1 \, dt)u_1}{dt} \tag{2.12}$$

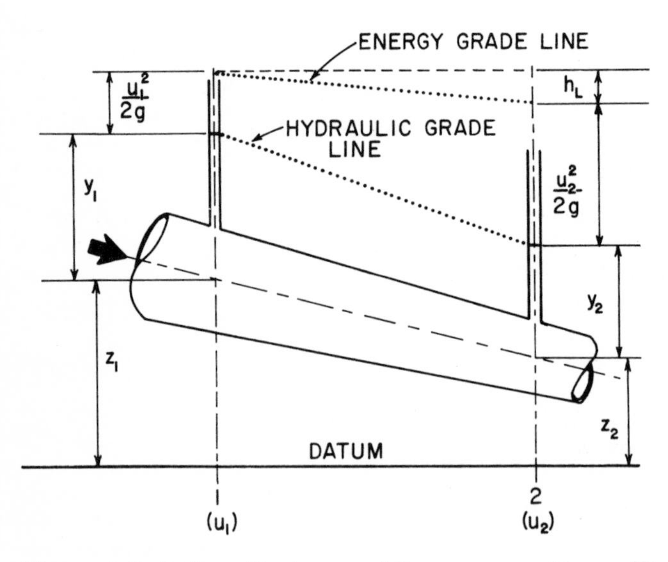

Figure 2.2 Defines the terms of the energy or Bernoulli equation.

where ΣF is the net force acting on the fluid body between points 1 and 2 in Figure 2.1; dt is an increment of time; and the term $\rho Au\,dt$ is the mass transferred across area A and hence $(\rho uA\,dt)u$ is the momentum across the cross-sectional area of flow A (subscripts 1 and 2 refer to the points shown in Figure 2.1).

2.3 FLOW EQUATIONS FOR LAMINAR FLOW

Shear Stress

Again consider our control volume of fluid flowing steadily in Figure 2.1. By performing a force balance on the system, the following is obtained:

$$\Sigma F = 0 = \rho Q(u_2 - u_1) = (P_1 - P_2)\pi R^2 - \tau_w(2\pi RL) \tag{2.13}$$

Rearranging and simplifying, this becomes

$$\tau_w = \frac{R}{2}\frac{\Delta P}{L} \tag{2.14}$$

where $\Delta P = P_1 - P_2$ and τ_w is the shear stress at the fluid-pipe wall interface.

A similar expression can be derived for the shear stress in the fluid, τ, at any arbitrary radius, r, such that by equating Eq. (2.14) with $\tau = (r/2)(\Delta P/L)$ we obtain:

$$\tau = \tau_w \frac{r}{R} \tag{2.15}$$

Equation (2.15) states that the shear stress varies linearly with pipe radius, as illustrated in Figure 2.3A.

Velocity Distribution

The velocity profile across the circular duct in Figure 2.1 can be derived by substituting our general expression for shear stress into Newton's law of viscosity [Eq. (1.3)] and integrating by applying the no-slip condition at the wall (i.e., at $r = R$, $u = 0$):

$$u = \frac{\Delta P}{4\mu L}(R^2 - r^2) \tag{2.16}$$

As described by Eq. (2.16) and illustrated in Figure 2.3B, the velocity profile of a fluid in laminar flow is parabolic.

Volumetric Flow Rate

Volumetric flow rate can be derived by performing the following integration:

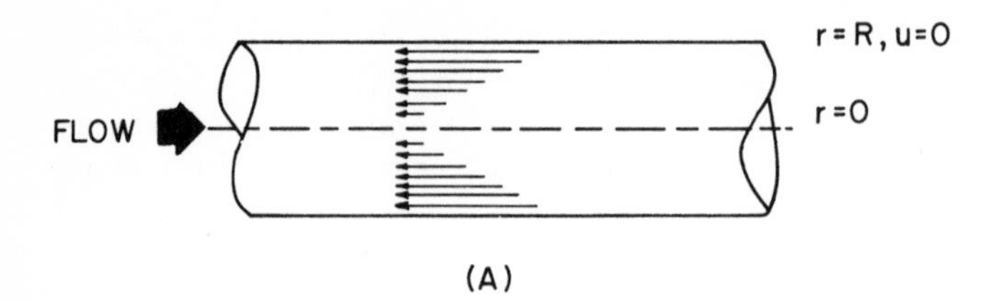

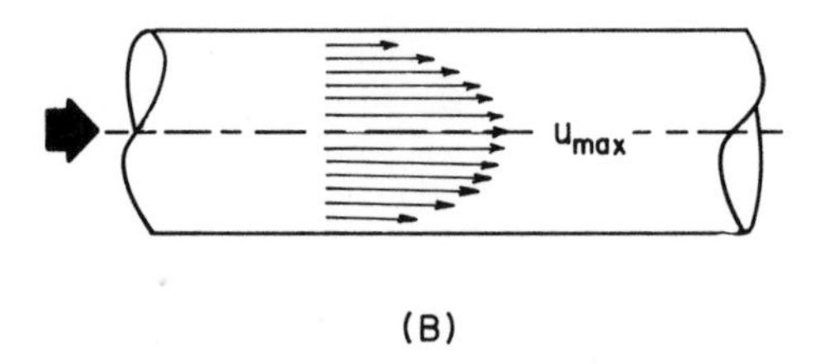

Figure 2.3 (A) The shear stress distribution in a fluid flowing in a circular tube. (B) The velocity profile of a fluid laminarly flowing in a circular tube.

$$Q = \int_0^R u(2\pi r\ dr) \tag{2.17}$$

where the expression for u is given by Eq. (2.16). Hence,

$$Q = \frac{\Delta P}{8\mu L}\ \pi R^4 \tag{2.18}$$

Equation (2.18) is known as the Hagen-Poiseuille relation.

Average Velocity

Average velocity is defined as

$$\bar{u} = \frac{Q}{A} = \frac{\Delta P}{8\mu L}\ R^2 \tag{2.19}$$

Nonisothermal Flow

The above expressions assume constant fluid properties (i.e., viscosity and density are constant). For isothermal, incompressible flow, this assumption is valid. However, if the fluid is undergoing heating or cooling, properties will vary.

For a heated tube with a fluid in laminar flow, the isothermal velocity profile becomes distorted and flattened.[10] This results from a radial temperature gradient in the fluid which causes the viscosity to lower in the vicinity of the wall. As such, the fluid will tend to flow at a greater velocity near the wall. (For a cooling situation, the viscosity of the fluid becomes greater near the wall resulting in a more pointed velocity profile.[11]

2.4 FLOW EQUATIONS FOR TURBULENT FLOW

Mixing Length

Turbulence can be described as the random, chaotic movement of fluid particles. Prandtl[12] derived the universal velocity profile describing the turbulent flow of Newtonian fluids in smooth tubes from mixing length theory. The mixing length equals the mean distance that an eddy travels in a direction normal to the main flow before losing its identity.

To derive the velocity distribution from Prandtl's theory, we consider the shear stress at a point in the fluid located at a distance y from the wall (see Figure 2.4) to be expressed by:

$$\text{shear at any point from wall} = \text{viscous forces shear} + \text{turbulent forces shear} \tag{2.20}$$

Mathematically this can be expressed as the following:

$$\tau = \frac{\mu}{g_c}\frac{du}{dy} + \frac{\rho}{g_c}\epsilon_M \frac{du}{dy} \tag{2.21}$$

where ϵ_M is the eddy diffusivity of momentum (also called eddy kinematic viscosity). The eddy diffusivity of momentum is assumed to be proportional to the mixing length (ℓ) and the mean eddy velocity (u_e).

$$\epsilon_M \propto \ell \frac{du}{dy} \tag{2.22}$$

and

$$u_e \propto \ell \frac{du}{dy} \tag{2.23}$$

hence

$$\epsilon_M \propto \ell^2 \frac{du}{dy} \tag{2.24}$$

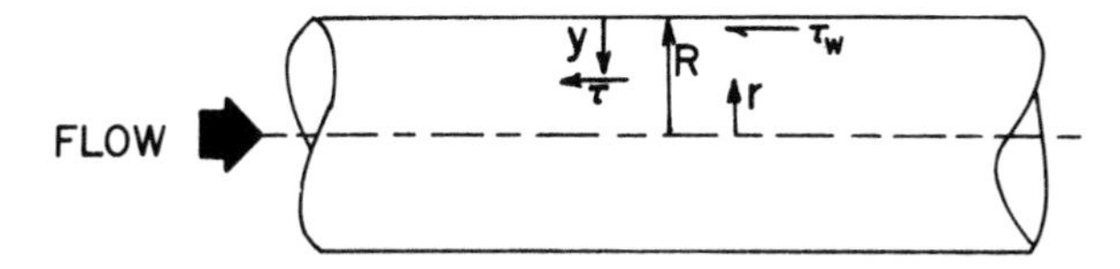

Figure 2.4 Orientation of the Prandtl mixing length derivation.

Substitution of Eq. (2.24) into Eq. (2.21) and assuming that viscous forces are negligible in the turbulent core results in the following expression:

$$\tau = \frac{\rho}{g_c} \ell^2 \left(\frac{du}{dy}\right)^2 \tag{2.25}$$

The mixing length is defined as

$$\ell = ky \tag{2.26}$$

where k is a proportionality constant.

From Eq. (2.15), which relates the shear stress at the wall to the shear at any distance from the wall, and from Figure 2.4, $r = R - y$. Then

$$\tau_w\left(1 - \frac{y}{R}\right) = \frac{\rho}{g_c}\left(ky\ \frac{du}{dy}\right)^2 \tag{2.27}$$

For regions near the wall $(1 - y/R)$ approaches unity and hence

$$\sqrt{\frac{\tau_w g_c}{\ell}} = ky\ \frac{du}{dy} \tag{2.28}$$

And introducing the friction velocity $u^* = \sqrt{\tau_w g_c/\rho}$

Rearranging and integrating we obtain

$$u = u^*\left(\frac{1}{k}\ \ln y + C'\right) \tag{2.29}$$

Evaluation of the constant of integration results in

$$\frac{u}{u^*} = \frac{1}{k}\ \ln\left(\frac{yu^*\rho}{k}\right) + k' \tag{2.30}$$

And we introduce the following definitions:

$$u^+ = u/u^* \qquad \text{(dimensionless velocity)} \tag{2.31}$$

$$y^+ = yu^*\rho/\mu \qquad \text{(dimensionless distance parameter)} \tag{2.32}$$

$$u^+ = \frac{1}{k}\ \ln y^+ + k' \tag{2.33}$$

Equation (2.33) is known as the universal velocity profile. The constants have been experimentally determined[13-15] to give the following:

For the laminar layer,

$$u^+ = y^+; \qquad y^+ < 5 \tag{2.34}$$

For the buffer layer,

$$u^+ = 5.0\ \ln y^+ - 3.05; \qquad 5 < y^+ < 30 \tag{2.35}$$

For the turbulent core,

$$u^+ = 2.5 \ln y^+ + 5.5; \qquad y^+ > 30 \tag{2.36}$$

These equations can be used to compute the complete velocity profile for a fluid in turbulent flow in a smooth pipe.

2.5 FRICTION FACTORS AND FRICTIONAL HEAD LOSSES

Darcy-Weisbach Equation

Equation (2.37) is the Darcy-Weisbach equation, which can be used to estimate the frictional head losses.

$$h_L = f \frac{L\bar{u}^2}{2gD} = \frac{\Delta P}{g\rho} \tag{2.37}$$

where

h_L = friction head loss (m)

f = friction factor coefficient

L = length of pipe (m)

D = inside pipe diameter (m)

$\bar{u}$ = average fluid velocity (m/s)

g = acceleration due to gravity ($981.4\ \text{cm/s}^2 = 9.81\ \text{m/s}^2$)

Note that if the flow is through a noncircular channel, Eq. (2.37) is expressed as:

$$h_L = f \frac{L\bar{u}^2}{8gR_H} \tag{2.38}$$

where R_H is the hydraulic radius, which is the cross-sectional area of flow divided by the wetted perimeter.

The friction factor f varies with the Reynolds number, Re, defined as

$$\text{Re} = \frac{\bar{u}D}{\nu} = \frac{DG}{\mu} = \frac{4W}{\pi D\mu} \tag{2.39}$$

and for noncircular ducts

$$\text{Re} = 4R_H G/\mu \tag{2.40}$$

where

$\nu = \mu/\rho$ = kinematic viscosity (cm^2/s)

G = mass flow per cross-section area ($\text{g/cm}^2 \cdot \text{s}$)

W = mass velocity (g/s)

and the other symbols are the same as previously defined.

The usual criteria for full pipe flow is that laminar conditions exist for Re < 2100, in which case the friction factor can be computed from the relationship

$$f = \frac{64}{Re} \tag{2.41}$$

For Reynolds numbers greater than 2100, the flow is considered turbulent.

The friction factor is a function of pipe roughness and size in addition to Reynolds number. The effect of pipe size and roughness is denoted by the ratio of the pipe's absolute roughness element, ϵ (in dimensions of feet), to pipe diameter, D (ft). Figure 2.5 shows the relationship between friction factor and ϵ/D and Re on the well-known Moody diagram.[16]

For highly turbulent conditions (i.e., large Re) and/or for large values of ϵ/D, the friction factor becomes less dependent on Re and in fact only

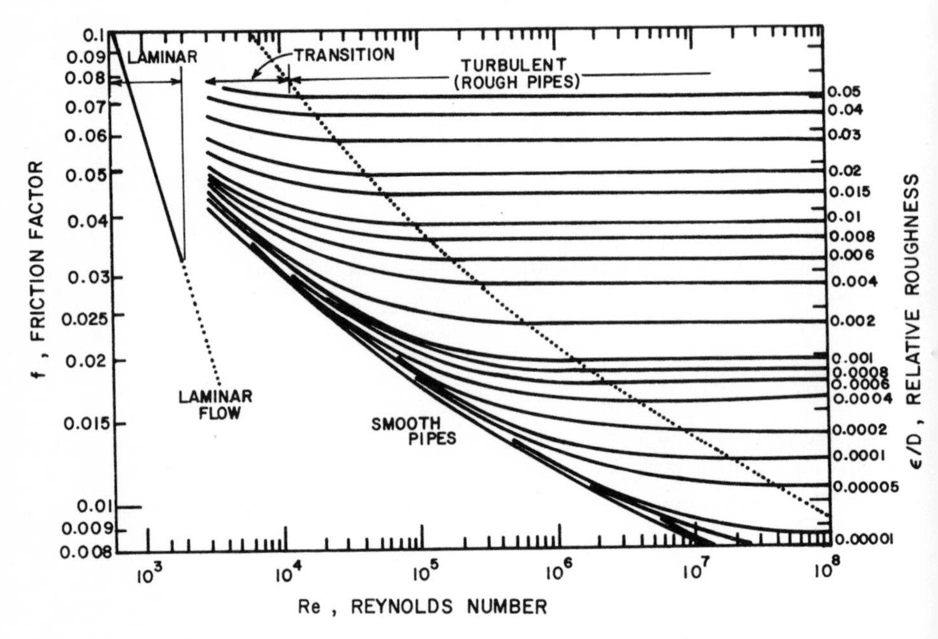

Figure 2.5 Moody diagram (see Ref. 16) for estimating friction factor, f, where

$$f = \frac{2h_L gD}{L\bar{u}^2}$$

becomes a function of ϵ/D. Under these conditions the resultant curves in Figure 2.5 are horizontal lines (the flow is often called wholly rough or rough flow). For a wholly turbulent (rough) condition the friction factor can either be estimated from Figure 2.5 or the following equation:

$$\frac{1}{\sqrt{f}} = 2 \log_{10} \frac{D}{\epsilon} + 1.14 \tag{2.42}$$

Equation (2.42) is considered applicable for Re > 4000.

Transition Region

Colebrook and White[17] recommend the following equation for the transition region between smooth and wholly turbulent flow:

$$\frac{1}{\sqrt{f}} = 1.14 - 2 \log_{10}\left(\frac{\epsilon}{D} + \frac{9.35}{Re\sqrt{f}}\right) \tag{2.43}$$

Manning Equation

The Manning equation [Eq. (2.44)] is an empirical expression originally developed in open channel flows; however, it can be applied to closed channel situations:

$$u = \frac{1.486}{\beta} R_H^{2/3} S^{1/2} \tag{2.44}$$

where u and R_H are in ft/s and ft, respectively, and S equals slope of the energy line, h_L/L. Noting that $u = Q/A$ and that for a circular pipe, $R_H = D/4$, the Manning expression can be expressed in terms of the head loss:

$$h_L = \frac{4.666 \beta^2 L}{D^{5.333}} Q^2 \tag{2.45}$$

where D and L are in feet, h_L is in feet of fluid and Q is in ft^3/s.

A range of typical values of the proportionality constant β for various pipe materials is given in Table 2.1.

Hazen-Williams Equation

The Hazen-Williams equation is given as follows:

$$u = 1.318 C R_H^{0.63} S^{0.54} \tag{2.46}$$

(again, English units). The proportionality constant C is often considered as a coefficient of roughness. Following the treatment applied to the Manning expression, Eq. (2.46) is expressed by the following:

Table 2.1
Range of β Values for Various Pipe Materials[18,19]

Pipe material	
Cast Iron	0.011 - 0.015
Wrought iron (black)	0.012 - 0.015
Wrought iron (galvanized)	0.013 - 0.017
Smooth brass	0.009 - 0.013
Glass	0.009 - 0.013
Riveted and spiral steel	0.013 - 0.017
Clay drainage tile	0.011 - 0.017
Concrete	0.012 - 0.016
Concrete lined	0.012 - 0.018
Concrete-rubble surface	0.017 - 0.030
Polyvinyl chloride (PVC)	0.008 - 0.012
Wood	0.010 - 0.013

Table 2.2
Typical Values of the Hazen-Williams Coefficient[19-21]

Pipe material	C
Very to highly smooth pipes (all metals)	130-140
Smooth wood	120
Smooth masonry	120
Vitrified clay	110
Cast iron (old)	100
Iron (worn/pitted)	60-80
PVC	150
Brick	100

$$h_L = \frac{4.73L}{C^{1.852}D^{4.87}} Q^{1.852} \tag{2.47}$$

with L and D in feet.

The equation is normally applied to channels discharging under pressure. Table 2.2 gives typical values of the Hazen-Williams coefficient C for various pipe materials.

2.6 ESTIMATING MINOR LOSSES

Entrance and End Effects

When a fluid enters a cylindrical pipe from a large reservoir, its velocity profile is flat. After a sufficient length of flow along the tube axis, fully developed flow results and the velocity profile follows the equations previously described.

The pressure gradient in the entrance region will be greater than at a distance downstream where the velocity profile has been fully developed. The amount of head loss in a pipe entrance largely depends on the geometry of the entrance. An estimate can be obtained from Eq. (2.48):

$$h_L = K_L \frac{u^2}{2g} \tag{2.48}$$

In general, the head loss resulting from added turbulence or secondary flows is roughly proportional to the flow rate or fluid velocity squared. Parameter K_L is referred to as the loss coefficient. Typical values for different entrance geometries are given in Figure 2.6.

The greater static pressure drop at the entrance region is attributed to two effects. These are the conversion of some kinetic pressure energy into kinetic energy as the core fluid is accelerated, and excessive fluid friction which is caused by the abnormally high velocity gradient in the region near the tube wall.

Exit head losses can be described similarly. If, for example, water discharges from a pipe into a large reservoir, its entire kinetic energy per pound of fluid dissipates within the reservoir.[19] Under these conditions the loss coefficient, K_L, in Eq. (2.48) equals unity.

Expansions

For expansions or enlargements, the velocity distribution behaves as shown in Figure 2.7. The minor head loss from this situation can be approximated from the following expression:

$$h_L = K_L \frac{(\bar{u}_2 - \bar{u}_3)^2}{2g} \tag{2.49}$$

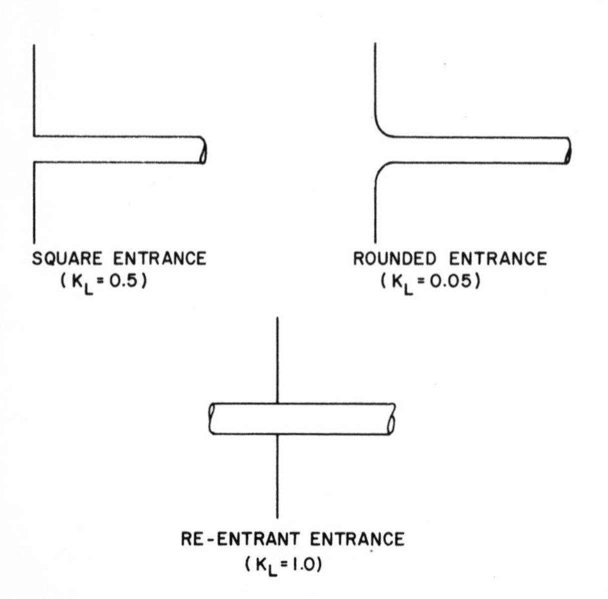

Figure 2.6 Various entrance geometries for a piping system.

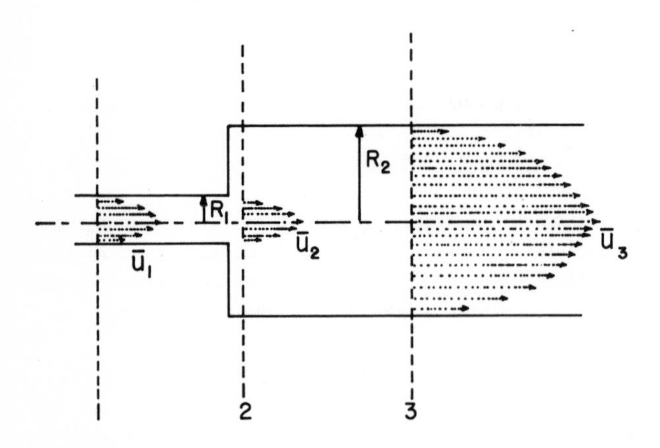

Figure 2.7 Sudden expansions in tubes.

where $\bar{u}_2$ is the average velocity in the smaller upstream pipe (it can be assumed that at the entrance $\bar{u}_1 = \bar{u}_2$) and $\bar{u}_3$ is the average velocity in the larger downstream channel. For the sudden expansion shown in Figure 2.7, $K_L = 1.0$. For gradual enlargements, K_L is a function of the interior angle of the enlargement. Archer[22] gives the following semiempirical relation for estimating losses due to sudden enlargements of closed pipes.

$$h_L = 1.098\left(\frac{\bar{u}_1}{2g}\right)^{1.919}\left(1 - \frac{A_1}{A_2}\right)^{1.919} \tag{2.50}$$

(English units).

Valves/Fittings/Bends

For valves and various fittings, minor head loss can largely be caused by separation of the flow lines from the boundary at abrupt changes of the wall geometry. Equation (2.48) can be used as a rough approximation for these losses. Table 2.3 gives values for the loss coefficient for various types of valves.

Slight changes in direction can produce disturbed flow conditions that increase the frictional resistance. The length in which such disturbances exists becomes more important than the sharpness of the deflection. For example, a curve of short radius (and correspondingly short length curve) tends to have a smaller head loss than a long radius curve undergoing the same change in direction. Table 2.4 gives nominal values for the loss coefficient in Eq. (2.48) for bends (fittings).

2.7 FLOW EQUATIONS FOR NON-NEWTONIAN FLUIDS

The flow behavior of non-Newtonian fluids is poorly understood, partly because of the lack of fundamental experimental studies. This subsection provides a brief overview of the models and equations that attempt to describe

Table 2.3
Loss Coefficients for Valves

Type valve	Position	K_L
Globe	Fully open	10
Angle	Fully open	5
Gate	Fully open	0.19
	Three-fourths open	1.0
	One-half open	5.6
Ball check	Fully open	70
Foot	Fully open	15
Swing check	Fully open	2.3

Table 2.4
Loss Coefficients for Fittings

Fitting	K_L
Close return bend	2.2
Standard short radius elbow	0.9
Medium sweep elbow	0.8
Long sweep elbow	0.6
45° elbow	0.4

the flow of time-independent fluids in a cylindrical tube under laminar and turbulent conditions.

In general turbulent flow studies of non-Newtonians are rare, primarily because turbulence is not as often encountered with these fluids in real-life situations as with Newtonians. Dodge and Metzner[23] present a theoretical and experimental study on non-Newtonian time-independent fluids. By use of the power law model they relate the Fanning friction factor, f, to a generalized Reynolds number (discussed in the next section).

Metzner[24] recommends using the relationships for newtonian fluids to estimate friction factors and head losses for non-Newtonian flows in rough tubes. Limited studies have shown that friction factors for pseudoplastics are smaller but not greatly different than values for Newtonian materials under comparable flow conditions.[2]

A number of semitheoretical velocity distributions for time-independent fluids have been developed by various investigators. These are summarized in Table 2.5.

Table 2.5
Semitheoretical Velocity Distributions

Principle model or study	Investigators
Power law fluids	Dodge and Metzner[23]
Experimental study of clay suspensions and carbopol solutions (viscoelastic fluids)	Boque and Metzner[25]
Power law fluids	Clapp[26]
Power law fluids	Pai[27]
Power law fluids	Brodkey et al.[28]

Laminar Flow

For fluids with a yield stress there are several empirical models that relate shear rate to shear stress (refer to Table 2.6). The Bingham plastic model[29] has been given the most analytical attention, although fluids following the model predictions are rare.

For a Bingham plastic, the apparent viscosity is defined as:

$$\mu_a = \eta + \frac{g_c \tau_y}{du/dy} \tag{2.51}$$

where τ_y is the fluid's yield stress (see Chapter 1) and η is the plastic viscosity.

From an analysis similar to the derivation of the Hagen-Poiseiulle relation (Eq. 2.18), the following expression can be proven:

$$\frac{Q}{\pi R^3} = \frac{8Q}{\pi D^3} = \frac{1}{4}\left(\frac{8\bar{u}}{D}\right) = \frac{g_c \tau_w}{4\eta}\left[1 - \frac{4}{3}\left(\frac{\tau_y}{\tau_w}\right) + \frac{1}{3}\left(\frac{\tau_y}{\tau_w}\right)^4\right] \tag{2.52}$$

Equation (2.52) is Buckingham's equation.[33] In the form given, the Buckingham equation cannot be directly solved for the frictional pressure loss ΔP; however, an approximate solution can be obtained by neglecting the term $1/3\ (\tau_y/\tau_w)^4$, which leaves:

$$\frac{Q}{\pi R^3} = \frac{8\bar{u}}{D} \simeq \frac{g_c}{\eta}\left[\tau_w - \frac{4}{3}\tau_y\right] \tag{2.53}$$

If a plot of $8\bar{u}/D$ versus τ_w is prepared, parameters η and τ_y can be obtained from the slope and intercept on the τ_w axis, respectively. Skelland[2] provides an example problem.

For the velocity distribution, Eq. (2.54) describes the general form for fluids with a yield stress:

$$\bar{u} = \int_0^u du = \int_r^R \left(-\frac{du}{dr}\right) dr \tag{2.54}$$

For Bingham plastics, Eq. (2.54) results in:

$$u = \frac{g_c}{\eta}\left[\frac{\Delta P}{L}(R^2 - r^2) - \tau_y (R - r)\right] \tag{2.55}$$

Table 2.6
Various Models Relating Shear Stress to Shear Rate (Fluids with a Yield Stress)[a]

Model	Form of equation for ($\tau > \tau_y$, du/dy = 0 for $\tau < \tau_y$)	Empirical constants	Units	References
Bingham plastic	$\tau - \tau_y = \frac{\eta}{g_c}\left(\frac{du}{dy}\right)$	τ_y	lb_f/ft^2	22
		η	$lb_m/ft \cdot s$	
Herschel-Bulkley	$\tau - \tau_y = \left[\frac{\eta'}{g_c}\left(\frac{du}{dy}\right)\right]^{1/m}$	τ_y	lb_f/ft^2	30, 31
		η'	$lb_f^{m-1} lb_m \cdot ft^{1-2m}/s$	
		m	dimensionless	
Crowley-Kitzes	$\tau = \frac{\mu_L}{g_c}\left[\frac{1.2 + \phi\,(c\tau^{-0.2} + 1)^3}{1.2 - 2\phi(c\tau^{-0.2} + 1)^3}\right]\left(\frac{du}{dy}\right)$	c	$(lb_f/ft^2)^{0.2}$	32

[a]Refer to Nomenclature section for definition of terms.

Fluids without a Yield Stress

As pointed out in Chapter 1, pseudoplastic fluids constitute the majority of non-Newtonian materials. For power law fluids, the following relation is applicable for flow through a tube:

$$\frac{Q}{\pi R^3} = \frac{n}{3n+1}\left(\frac{\tau_w g_c}{K}\right)^{1/n} \tag{2.56}$$

and by rearranging Eq. (2.56) the pressure gradient can be obtained from:

$$\frac{\Delta P}{L} = 2K\left(\frac{3n+1}{n}\right)^n \frac{u^n}{g_c R^{n+1}} \tag{2.57}$$

Note that for highly pseudoplastic materials the flow behavior index n approaches zero.

To obtain the velocity distribution of a power law fluid in laminar flow in a tube, these expressions are rearranged to give the following:

$$u = \left(\frac{\Delta P g_c}{2KL}\right)^{1/n} \frac{n}{n+1}\left(R^{(n+1)/n} - r^{(n+1)/n}\right) \tag{2.58}$$

which is analogous to Eq. (2.16); or

$$u = \bar{u}\left(\frac{3n+1}{n+1}\right)\left[1 - \left(\frac{r}{R}\right)^{(n+1)/n}\right] \tag{2.59}$$

where u is the local velocity and $\bar{u}$ is the average velocity. Velocity profiles for different values of n were computed from Eq. (2.59) and are shown in the plot in Figure 2.8. Note that for very low n values, the flow approaches plug.

The power law model can also be used to estimate a friction factor:

$$f = \frac{\tau_w g_c}{\rho \bar{u}^2/2} = 16\left[\frac{D^n \bar{u}^{2-n} \rho}{K} \, 8\left(\frac{n}{6n+2}\right)^n\right]^{-1} = 16/\text{Re}' \tag{2.60}$$

Here Re' is defined as an effective Reynolds number for tube flow of power law fluids where

$$\text{Re}' = \frac{Du\rho}{\mu'} \tag{2.61}$$

μ' is referred to as the effective viscosity, which is the viscosity that allows Poiseuille's equation to fit any set of laminar flow conditions for time-independent fluids:

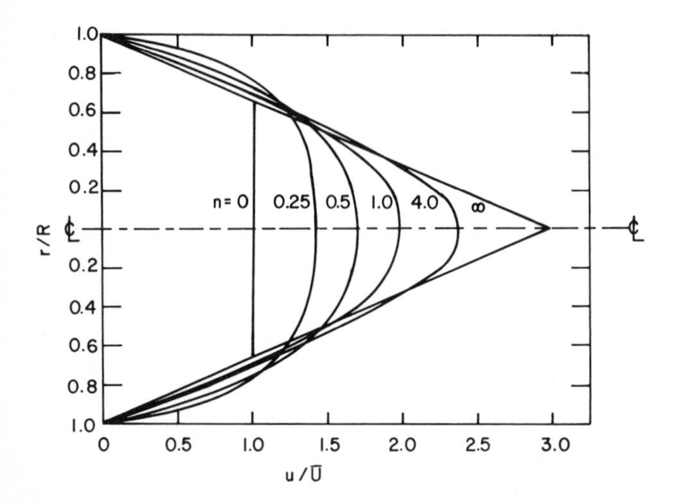

Figure 2.8 Velocity distributions for power law fluids in laminar flow through a cylindrical tube.

$$\mu' = \frac{D\,\Delta P\, g_c/L}{8\bar{u}/D} = \frac{g_c \tau_w}{8\bar{u}/D} \tag{2.62}$$

and for a Bingham plastic fluid the effective viscosity is defined by:

$$\mu' = \frac{\tau_y g_c D}{6\bar{u}} + \eta \tag{2.63}$$

Nonisothermal Laminar Flow

In Chapter 1 it was pointed out that the viscosity of an incompressible fluid will decrease with temperature. The viscosity-temperature relationship for Newtonian fluids is described by the Arrhenius equation:

$$\mu = ae^{E/R_G T} \tag{2.64}$$

where

E = activation energy of flow per mole

R_G = is the gas constant per mole

T = is absolute temperature

a = constant characteristic of the fluid

For a power law fluid this can be expressed in terms of the shear stress as follows:

$$\tau g_c = k_\sigma \left[\left(-\frac{du}{dr} \right) \exp\left(\frac{E}{R_G T} \right) \right]^n = k_\sigma S_r^n \tag{2.65}$$

where S_r is referred to as the reduced shear rate and k_σ, E, and n are constants independent of temperature.

2.8 OPEN CHANNEL FLOW

Open channel flow can be defined as the flow of fluids in a channel whose geometry is such that one liquid surface is free of solid boundaries.

Earlier, the term hydraulic grade line was introduced. In full pipe flow this term refers to the line that connects the points to which liquid would rise at various stations along the pipe if piezometers were positioned in the liquid. That is, it is a measure of the achievable pressure head at these various stations. In open channel flow the hydraulic grade line corresponds to the profile of the liquid surface.

Consider liquid flowing in a horizontal-bottomed open channel. The specific-energy concept can be employed to analyze the flow. Specific energy (E_s, sometimes called the specific head) is the sum of the pressure head and the velocity head $\bar{u}^2/2g$ measured with respect to the channel bottom. The relationship between specific energy and depth of flow, for a constant volumetric flow rate, is shown in Figure 2.9.

The specific-energy diagram indicates that flow in two regimes is possible for a specified specific energy. At the minimum energy ($E_{s1\ min}$), the liquid depth is referred to as the critical depth, δ_c. If the actual depth

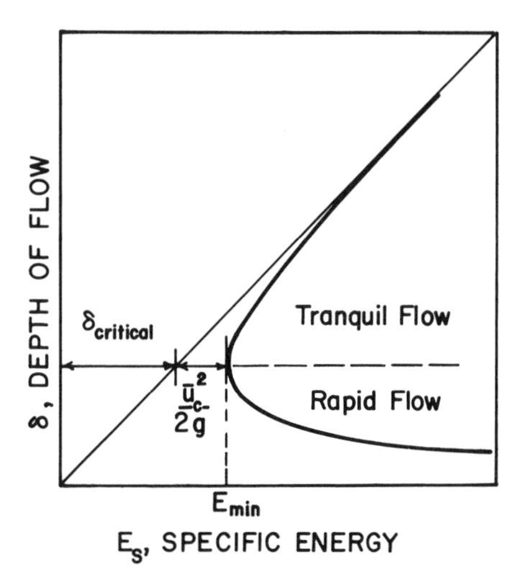

Figure 2.9 The specific-energy diagram.

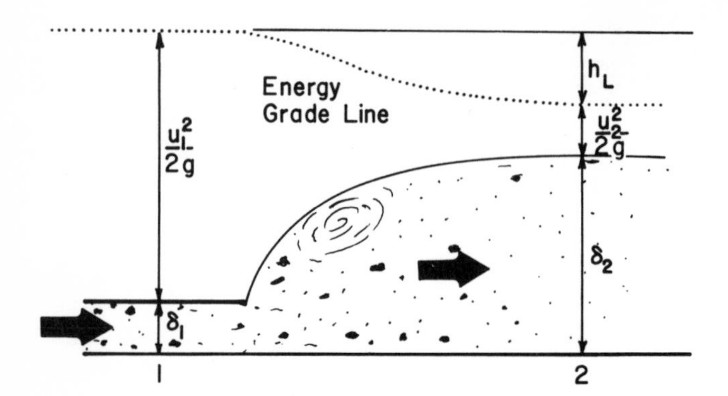

Figure 2.10 Illustrates a hydraulic jump.

is greater than δ_c, the flow is tranquil; however, if it is less than δ_c, the flow is considered rapid. Aside from outside influence, the flow may change abruptly with large losses from the rapid to the tranquil regime by means of a hydraulic jump (refer to Figure 2.10).

If the flow in Figure 2.10 is considered to be two dimensional and we neglect friction, the momentum equation [Eq. (2.12)] can be applied to give:

$$\Sigma F = F_1 - F_2 = \frac{\rho\delta_1^2}{2} - \frac{\rho\delta_2^2}{2} = \frac{Q\delta(\bar{u}_2 - \bar{u}_1)}{2} \tag{2.66}$$

For a given flow rate per unit channel depth q, where by continuity

$$q = \bar{u}_2\delta_2 = \bar{u}_1\delta_1 \tag{2.67}$$

Substitution of $\bar{u}_1$ and $\bar{u}_2$ into Eq. (2.66) results in the following expression, which can be used to compute the various depths of flow in Figure 2.10:

$$\frac{\delta_1}{\delta_2} = 1/2\left[\sqrt{\frac{8q^2}{gy_2^3}} - 1\right] \tag{2.68}$$

Flow problems in open channels can be tackled with the Manning equation [Eq. (2.45)].

2.9 TYPES OF FLOW MEASUREMENT

Fluid flow measurement is necessary to a wide range of applications. Numerous research projects and industrial processes rely on measurements to furnish data for analysis. Under laboratory conditions, high precision may be necessary. In industrial practice, accurate measurements may be necessary for the testing and acceptance of materials-handling equipment. Less precise measurements may be used for material balances around various separation processes such as distillation, extraction, polymerization, or in wastewater-handling processes such as sewage treatment. Whatever the application, it is essential that all variables be considered in selecting the proper flow measuring technique. Variables that should be given consideration are meter accuracy, maintenance, type of indication and flow control unit, flexibility, resistance to the materials being processed, properties of the fluid, and economic viability.

The major classes of flow meters and flow measuring techniques for gas and liquids to be covered in the chapters that follow are outlined below:

A. Differential pressure meters
   1. Venturi
   2. Flow nozzles
      2.1 Critical flow nozzle
   3. Orifice
   4. Pitot tube
      4.1 Impact tube
      4.2 Pitot-static tube
      4.3 Pitot-Venturi
B. Positive displacement
   1. Reciprocating piston
   2. Nutating disk
   3. Rotary piston
   4. Rotary vane
C. Mechanical
   1. Rotameter
      1.1 Glass
      1.2 Metal
   2. Turbine meter
      2.1 Electromagnetic
D. Acoustic
   1. Travel time difference method
   2. Beam deflection method
   3. Doppler method
E. Thermal
   1. Hot wire
   2. Hot film
F. Open channel techniques
   1. Rectangular weirs
   2. V-Notch weirs
   3. Other

NOMENCLATURE

A	Area (m^2, ft^2)
a	Constant in Eq. (2.64)
C	Coefficient of roughness, Eq. (2.46)
c	Constant in equation in Table 2.6 $(lb_f/ft^2)^{0.2}$
D	Diameter (m, ft)
E	Sum of energies in Eq. (2.3) (N·m/kg or lb·ft/slug)
E	Activation energy, Eqs. (2.64) and (2.65) (ft·lb_f/lb·mol)
E_k	Kinetic energy (N·m/kg or lb·ft/slug)
E_h	Lost energy per mass fluid, Eq. (2.7) (ft·lb_f/lb_m)
E'_h, E'_m	Defined in Eq. (2.9) (lb·s^2/slug)
E_m	Mechanical energy (ft·lb_f/lb_m)
E_p	Potential energy (N·m/kg or lb·ft/slug)
E_p	Pressure energy (N·m/kg or lb·ft/slug)
E_s	Specific energy, see Figure 2.9 (N·m or lb·ft)
F	Force (N or lb_f)
f	Friction factor
G	Mass flow rate per cross-sectional area (g/cm^2·s or lb/ft^2·s)
g	Acceleration of gravity (981.4 cm^2 or 32.2 ft/s^2)
g_c	Conversion factor (32.174 lb_m·ft/$lb_f s^2$)
H_P	Hydraulic head (m or ft of fluid)
H_T	Stagnation head (m or ft of fluid)
h_L	Friction head loss (m or ft of fluid)
K	Proportionality constant in Eq. 2.57
K_L	Proportionality constant in Eqs. (2.48) and (2.49)
k	Proportionality constant in Eq. (2.26)
k'	Constant in Eq. (2.33)
k_σ	Constant in Eq. (2.65)
L	Length (m or ft)
ℓ	Mixing length (m or ft)
m	Constant in equation in Table 2.6
n	Flow behavior index
P	Pressure (N/m^2 or lb_f/ft^2)
Q	Volumetric flow rate (m^3/s or ft^3/s)
q	Flow rate per unit channel depth (m^2/s, ft^2/s)
R	Pipe radius (m or ft)
R_G	Gas constant (ft·lb_f/lb·mol·°R)
R_H	Hydraulic radius (m or ft)
Re	Reynolds number, Eq. (2.39)
Re'	Effective Reynolds number, Eq. (2.61)
r	Radius (m or ft)
S	Slope of energy line, Eq. (2.44)
S_r	Reduce shear rate, defined in Eq. (2.65)
T	Temperature (K or °R)

t	Time (s)
u	Velocity (m/s or ft/s)
$\bar{u}$	Average velocity (m/s or ft/s)
u_e	Mean eddy velocity (m/s or ft/s)
u^*	Friction velocity, $\sqrt{\tau_w g_c/\rho}$ (m/s, ft/s)
u^+	Dimensionless velocity, Eq. (2.31)
W	Mass velocity (g/s or lb/s)
y	Distance from wall (m or ft)
y^+	Dimensionless distance, Eq. (2.32)
z	Vertical distance above datum (m or ft)

Greek Letters

β	Proportionality constant in Eqs. (2.44) and (2.45)
δ	Fluid depth (m or ft)
ϵ	Roughness element (ft)
ϵ_M	Eddy diffusivity of momentum (cm^2/s or ft^2/s)
η	Plastic viscosity (centipoise, cP)
η'	Constant in equation in Table 2.6 ($lb_f^{m-1} \cdot lb_m \cdot ft^{1-2m}/s$)
μ	Viscosity (cP)
μ'	Effective viscosity, Eqs. (2.62) and (2.63) (cP)
ν	Kinematic viscosity (μ/ρ) (cm^2/s or ft^2/s)
ρ	Density (g/cm^3 or lb/ft^3)
τ	Shear stress (N/m^2 or lb_f/ft^2)
τ_w	Wall shear stress (N/m^2 or lb_f/ft^2)
τ_y	Yield stress (N/m^2 or lb_f/ft^2)
ϕ	Volume fraction of solids in suspension

PROBLEMS

2.1 Four pipes join at a common junction as shown in Figure 2.11. Pipe diameters and flow rates in pipes 1, 3, and 4 are given in the figure. Using the conservation of mass principle, (a) compute the volumetric flow rate in pipe 2 and (b) compute the average fluid velocity in each pipe.

2.2 A 3-in. (0.0762-m) i.d. tube discharges water into a 7.5-in. (0.1905-m) tube. The average velocity in the large diameter pipe is 15 m/s. Compute the following:
(a) the volumetric flow rate;
(b) the mass flow rate;
(c) the average velocity in the 3-in. tube;
(d) the Reynolds number (μ = 1.09 cP at 20°C).

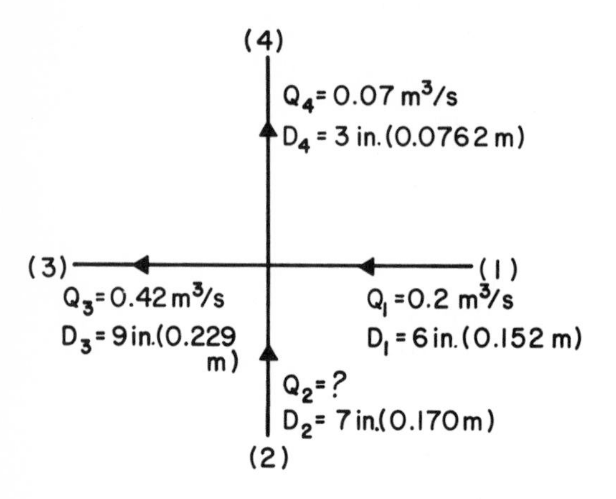

Figure 2.11

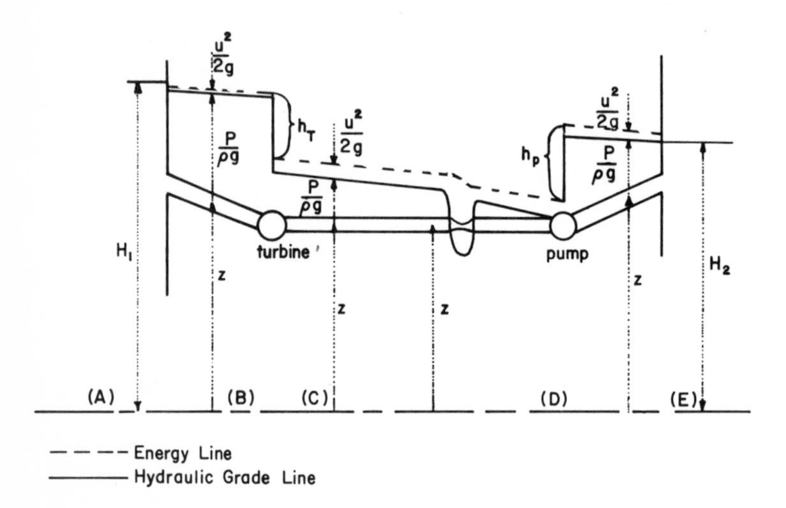

Figure 2.12

2.3 For Figure 2.12, write the Bernoulli equation between:
(a) sections A and B;
(b) sections A and C;
(c) sections D and E.

2.4 Water is flowing at a rate of 0.02 m^3/s through a 5-in. i.d. (12.7-cm) pipeline, roughly 70 m in length. Determine (a) the head loss per unit length of pipe and (b) the frictional pressure drop across the length of pipe.

2.5 For Problem 2.4, calculate the fluid's wall shear stress.

2.6 The rheological properties of a liquid were studied and found to closely fit the power law model. Fluid density and the power law constants obtained from curve-fitting viscosity data at the average process temperature are tabulated below. The fluid is flowing at a mass rate of 5200 lb/h in a 1-in. i.d. tube. Calculate the frictional pressure drop over a 10-ft straight length of pipe.

Flow behavior parameters:

K ($lb_m \cdot s^{n-2} \cdot ft^{-1}$)	n	ρ (lb_m/ft^3)	T (°F)
22	0.75	67	225

2.7 For Problem 2.6, compute:
(a) the centerline velocity of the fluid;
(b) the centerline velocity of the fluid in a 0.5-, a 2.0-, and a 3.0-in. tube.

2.8 Water is flowing through a 2-in. i.d. pipeline (smooth) 300 ft in length. The line contains a gate valve, one-half open, and the total head loss is known to be 18 ft.
(a) Compute the average fluid velocity and volumetric flow rate that can be expected,
(b) What percentage of the total head loss does the valve contribute?

3

Flow Measurement by Differential Flow Techniques

3.1 INTRODUCTION

Differential pressure flow measuring devices are generally termed head meters or rate meters. Their primary characteristics are that they measure the flow rate without sectioning the fluid into isolated quantities, and its primary device creates a differential head which is a function of the fluid velocity and density.

There are numerous examples throughout ancient history where the head meter has been applied. In ancient Rome head meters were used to measure water to householders and later the hourglass was applied to measuring time.

In the seventeenth century, Benedetto Castelli and Evangelista Torricelli established the groundwork theory for the modern day head meter. In the eighteenth century John Bernoulli established the principle on which the hydraulic equations for head meters are based (see Chapter 2), and in the latter part of that century Giovanni Battista Venturi published his famous work on the principles of the venturi tube.

There are a variety of devices that are based on the differential head method. In this chapter some of the more commonly used devices are discussed. These systems include venturi meters, flow nozzles, orifice meters, and pitot tubes.

3.2 THE VENTURI METER

In 1797, Venturi presented his work on the venturi tube;[34] however, it was not until 1887 that his principles were put to use when Clemens Herschel[35] fabricated the first commercial venturi tube.

Figure 3.1 illustrates the basic design of a venturi meter. As fluid passes through the reduced area of the venturi throat, its velocity increases,

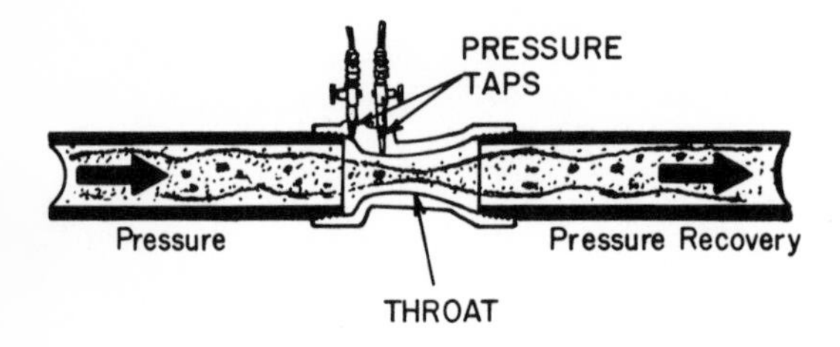

Figure 3.1 Illustrates operation of the venturi meter.

resulting in a pressure differential between the inlet and throat regions. That portion immediately following the throat gradually increases in flow area; consequently the fluid's velocity decreases, causing pressure recovery. The differential pressure across the venturi's throat can be recorded directly or translated into actual flow units [e.g., gallons per minute (GPM) or liters per second] by employing various types of differential pressure meters and capacity curves.

Venturi tubes are employed whenever the permanent loss of pressure is to be reduced to a minimum or in cases where fluids handled contain sufficient amounts of materials in suspension that other devices such as orifice plates or flow nozzles are not effective. These systems are recommended for use where metering conditions require relatively low pressure loss. Widest application has traditionally been in low-pressure gas lines and water mains. As noted above, its streamlined design makes the venturi suitable for metering liquids with solids in suspension. Venturi tubes are employed in hydronic systems that use hot and chilled water for heating and air conditioning. They are used in office buildings, manufacturing plants, hotels, or other edifices where comfort is an important criterion. Units are used in the processing industries where piping systems transport chemicals.

In many fan systems, venturi tubes are positioned in return lines for the purpose of reheating coils. Usually a balancing valve is installed downstream to allow accurate balancing.[36]

Another typical application is illustrated in Figure 3.2, in which a venturi with fixed meter is installed in a bypass to flow-test fire pump systems.

Venturi meters have long been used for measuring sewage flow. At each piezometer or annular chamber on the units, valves are positioned so that pressure openings can be closed. The valves are designed so that, upon closing, a rod is forced through the opening to clean out any debris that clogs. When all the valves are closed, plates covering the hand holes in the pressure chamber can be dismantled and the chamber flushed.

Design Considerations

Whenever a finite length exists between two pressure taps, the differential developed is the sum of the energy-transfer effect and the frictional effect. For a venturi tube there is an appreciable distance between the taps and a persistent decrease in diameter. Both these factors exert a significant impact on the differential developed in cases where fluid-friction values are high. For solids in suspension, fluid friction and the differential pressure are sensitive to the solids concentration.

The American Society of Mechanical Engineers (ASME)[37,38] has standardized the construction of these type of head meters. Figure 3.3 provides the ASME-recommended specifications for venturi meters. The design calls for pressure taps to be connected to manifolds which encompass the upstream and throat portions of the tube. The manifolds receive a portion of the pressure all around the sections so that an average differential can

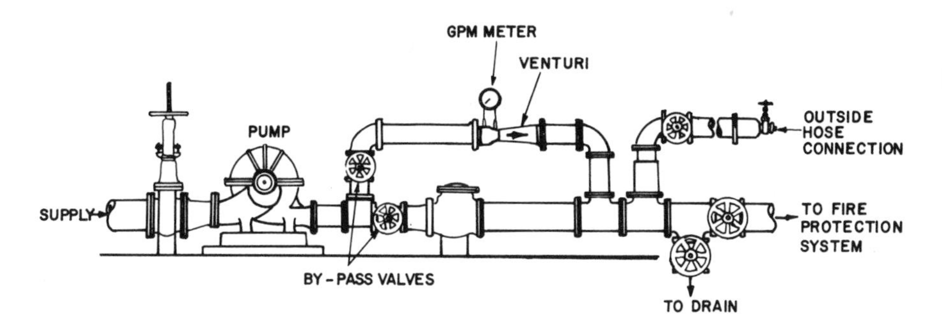

Figure 3.2 Venturi meter application to fire pump flow testing. (Courtesy of Aeroquip Corp., subsidiary of Libbey-Owens-Ford Co., Jackson, Mich.)

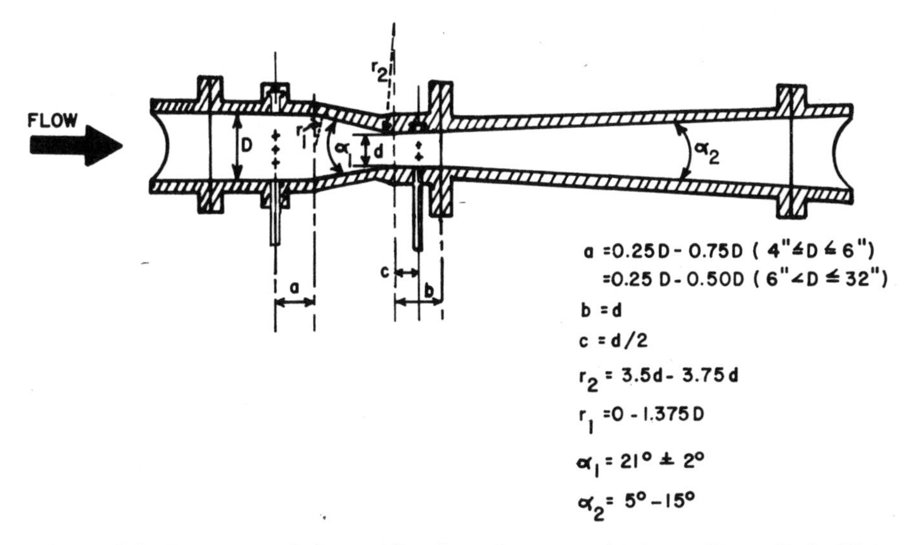

Figure 3.3 Recommended specifications for venturi tubes. (From Ref. 37.)

be obtained. A piezometer ring is inserted at the large end of the inlet section, and the determination of fluid flowing is based on the difference in pressures indicated at this point and at the throat.

In general, variations from these specifications in the outlet cone have a small effect on the discharge coefficient, but they do impact on the pressure recovery. The outlet cone may, however, be truncated considerably on low diameter ratios without seriously impacting on the pressure recovery (refer to Figure 3.4). In applications where there are space limitations a truncated cone venturi tube is to be preferred over the full-length exit cone.

Several modified designs are also used. Figure 3.5 illustrates two of them.

Incompressible Flow

If the fluid is incompressible and the flow is adiabatic and frictionless, then the Bernoulli expression [Eq. (2.9)] can be written as follows:

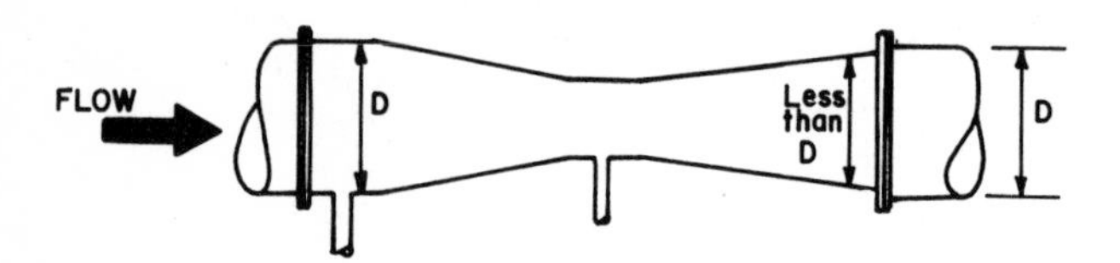

Figure 3.4 Illustrates truncated exit cone venturi tube.

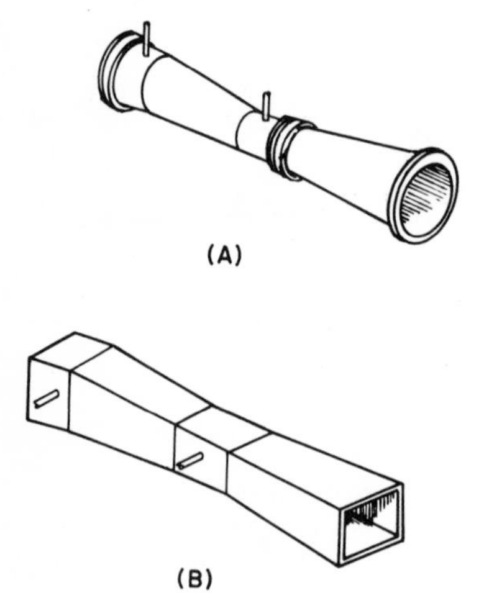

Figure 3.5 (A) Eccentric venturi tube. (B) Rectangular design venturi tube.

$$\frac{P_1}{\rho_1} + \frac{u_1^2}{2g_c} = \frac{P_2}{\rho_2} + \frac{u_2^2}{2g_c} \tag{3.1}$$

where subscript 1 refers to the inlet of the venturi and 2 refers to the throat. Combining Eq. (3.1) with the continuity equation (2.1), with $\rho_1 = \rho_2$, an expression for the pressure drop can be derived:

$$\Delta P = \frac{u_2^2 \rho}{2g_c}\left[1 - \left(\frac{A_2}{A_1}\right)^2\right] \tag{3.2}$$

and for the volumetric flow rate

$$Q_i = u_2 A_2 = u_1 A_1 = \frac{A_2}{\sqrt{1 - (A_2/A_1)^2}} \sqrt{\frac{2g_c}{\rho} \Delta P} \tag{3.3}$$

Equation (3.3) gives the theoretical or ideal volumetric flow rate. The ratio of the actual volumetric flow rate to the ideal value is known as the discharge coefficient.

$$C = \frac{Q_a}{Q_i} \tag{3.4}$$

It should be noted that the empirical coefficient may not be constant but rather a strong function of the channel geometry and Reynolds number.

The following semiempirical equation can be used to compute the actual volumetric flow rate for a venturi.

$$Q_a = CMA_2 \sqrt{\frac{2g_c}{\rho} (P_1 - P_2)} \tag{3.5}$$

M is referred to as the velocity of approach factor. This empirical constant is defined by Eq. (3.6):

$$M = \left[1 - \left(\frac{A_2}{A_1}\right)^2\right]^{-1/2} \tag{3.6}$$

Compressible Flow

Consider the case of an ideal gas, whereby the equation of state applies:

$$P = \rho RT \tag{3.7}$$

If the flow is reversible adiabatic, the steady-flow energy equation can be written as

$$C_P T_1 + \frac{u_1^2}{2g_c} = C_P T_2 + \frac{u_2^2}{2g_c} \tag{3.8}$$

where C_P is the specific heat of the fluid (for an ideal gas, C_P is constant). Combining the above expressions with the continuity equation gives:

$$\dot{m}_i^2 = 2g_c A_2^2 \frac{\gamma}{\gamma - 1} \frac{P_1^2}{RT_1} \left[\left(\frac{P_2}{P_1} \right)^{2/\gamma} - \left(\frac{P_2}{P_1} \right)^{(\gamma+1)/\gamma} \right] \tag{3.9}$$

where

$\dot{m}_i = \rho Q_i$ = the ideal mass flow rate (kg/s)

γ = ratio of the specific heat of the gas at constant pressure (C_P) to the specific heat at constant volume (C_V)

Equation (3.9) can be simplified (for the case where $\Delta P < P_1/10$) to the following expression for the actual mass flow rate ($\dot{m}_a$):

$$m_a = YCMA_2 \sqrt{2g_c \rho_1 (P_1 - P_2)} \tag{3.10}$$

the additional parameter Y is known as the expansion factor defined by Eq. (3.11)[39]:

$$Y = \left[\left(\frac{P_2}{P_1} \right)^{2/\gamma} \frac{\gamma}{\gamma - 1} \frac{1 - (P_2/P_1)^{(\gamma-1)/\gamma}}{1 - (P_2/P_1)} \frac{1 - \beta^4}{1 - \beta^4 (P_2/P_1)^{2/\gamma}} \right]^{1/2} \tag{3.11}$$

The parameter β is the diameter ratio (i.e., the venturi's throat diameter d divided by the inlet diameter D):

$$\beta = \frac{d}{D} = \sqrt{\frac{A_2}{A_1}} \tag{3.12}$$

Design and Installation Considerations

In specifying a venturi for a particular application, the first step should be to determine the flow requirements and establish all line sizes in the piping system. The semitheoretical formulas just presented can be used to size the units, or more often the manufacturer's capacity curves will allow fast estimates. Figure 3.6 shows a typical capacity curve for one manufacturer's design. Venturis are normally available in a range of β ratios for each line size.

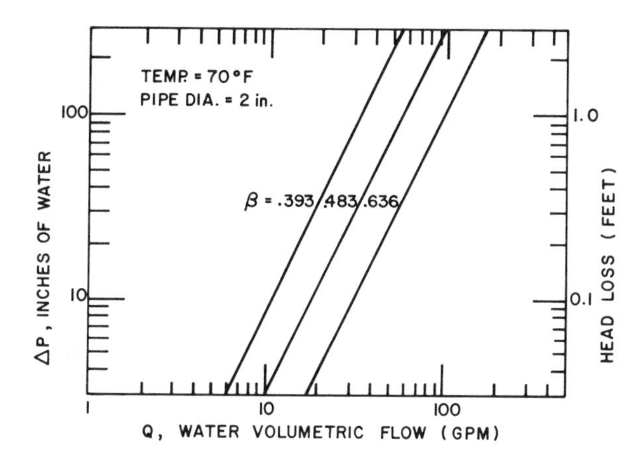

Figure 3.6 Manufacturer's flow curves for venturi in a 2-in. pipeline. (Courtesy of Aeroquip Corp., subsidiary of Libbey-Owens-Ford Co., Jackson, Mich.)

Each β ratio will have a separate capacity or flow curve. As shown in Figure 3.6, capacity curves provide information on the differential pressure at various volumetric flow rates. Note that since the plot is on the log-log scale, the flow curve is linear.

Temperature corrections and conversion factors for fluids other than water are usually provided with the capacity curves such as Figure 3.6. Figure 3.7 provides volumetric flow and differential pressure corrections for water flows at elevated temperatures.

For other fluids, capacities can be converted to equivalent GPM water at 70° F by the following set of equations.[36]

For liquids other than water,

$$Q = Q_f \sqrt{S_g} \tag{3.13}$$

where

Q is the sizing quantity equivalent GPM at 70° F, water

Q_f is given flow of fluid

S_g is the fluid's specific gravity

Note that there are no specific guidelines for viscous fluids; normally the manufacturer must be consulted for recommendations.

For air flow, an Aeroquip publication[36] recommends

$$Q = \frac{Q_f}{C_A} F_{pa} F_{ta} \tag{3.14}$$

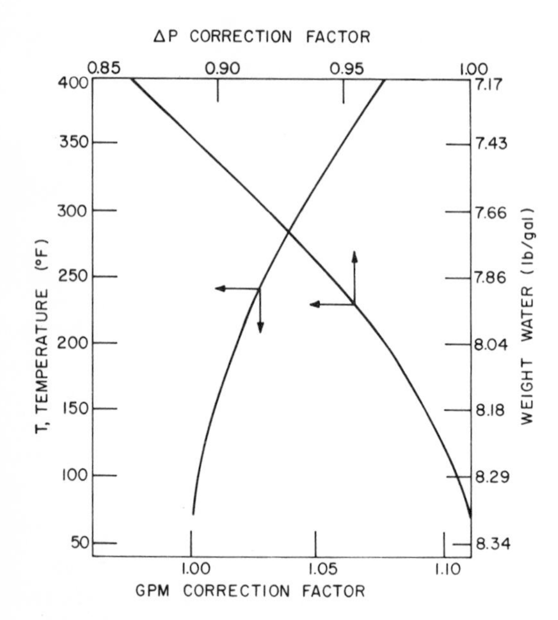

Figure 3.7 Differential pressure and volumetric flow corrections for water at elevated temperatures. To use, multiply ΔP specified for flow rate with 70° F water by the correction factor to obtain the ΔP for a specified flow at another temperature. To obtain the corrected volumetric flow, multiply GPM by the correlation factor to obtain the actual GPM at the indicated temperature. (Courtesy of Aeroquip Corp., subsidiary of Libbey-Owens-Ford Co., Jackson, Mich.)

where C_A = 3.8 SCFM at 0 psig and 70° F (equivalent to 1.0 GPM of water at 70° F); F_{pa} is the pressure correction factor for air; and F_{ta} is the temperature correction factor for air. Values for F_{pa} and F_{ta} are given in Table 3.1.

For saturated steam flow,

$$Q = \frac{Q_f}{C_s} F_{ps} \tag{3.15}$$

where C_s = 12.25 lb/h at 0 psig saturated (212°F), equivalent to 1.0 GPM of water at 70° F, and F_{ps} is the pressure correction factor (see Table 3.1). For gas flow other than air, Eq. (3.14) is applicable for rough approximations with $C_A = 3.8/\sqrt{S_g}$ SCFM at 0 psig and 70° F.

Installing a venturi consists of setting the unit into the line as another section of the pipe. The shorter cone forms the inlet or upstream end. The meter can be installed in the horizontal, vertical, or inclined position.

Table 3.1
Recommended Values for Pressure and Temperature Correction Factors for Air and Steam Flow[a]

Temperature		Pressure		
°F	Air/gas temp. (F_{ta})	psig	Air/gas pressure (F_{pa})	Sat. steam pressure (F_{ps})
0	0.932	0	1.000	1.000
2	0.933	2	0.938	0.934
4	0.936	4	0.886	0.887
6	0.938	6	0.843	0.846
8	0.940	8	0.805	0.811
10	0.942	10	0.771	0.780
12	0.944	12	0.742	0.752
14	0.946	14	0.716	0.727
16	0.948	16	0.692	0.705
18	0.950	18	0.670	0.685
20	0.952	20	0.651	0.666
25	0.956	25	0.608	0.626
30	0.961	30	0.573	0.592
35	0.966	35	0.544	0.564
40	0.971	40	0.518	0.539
50	0.981	50	0.477	0.498
60	0.990	60	0.443	0.466
70	1.000	70	0.416	0.439
80	1.009	80	0.394	0.416
90	1.019	90	0.375	0.397
100	1.028	100	0.358	0.380
120	1.046	120	0.330	0.352
140	1.064	140	0.308	0.331
160	1.081	160	0.290	0.312
180	1.099	180	0.275	0.296
200	1.116	200	0.261	0.282
225	1.137	225	0.247	0.267
250	1.157	250	0.235	0.255
275	1.177	275	0.225	0.244
300	1.197	300	0.216	0.234
325	1.217	325	0.208	0.226
350	1.236	350	0.201	0.218
375	1.255	375	0.194	0.211
400	1.274	400	0.188	0.204
425	1.292	425	0.183	0.198
450	1.310	450	0.178	0.193
475	1.328	475	0.173	0.188
500	1.346	500	0.169	0.183

[a]Courtesy of Aeroquip Corp., subsidiary of Libbey-Owens-Ford Co., Jackson, Mich.

It is good practice to install the venturi as far downstream as possible from the source(s) of flow disturbances (e.g., reducers, valves, or combinations of fittings). Figure 3.8 illustrates the proper hookups of a venturi for liquid and gas flow.

When the venturi is installed in a vertical line, pressure connections can obviously be made to any part of the tube. In horizontal or inclined geometries, care should be taken to ensure that pressure connections are installed in proper locations, or else faulty measurements will result.

For gas flow applications, locate pressure connections at the top of the tube.

For liquid flow applications, make connections at the side of the tube.

For steam installations where the meter is above the line, connections should be made at the top of the tube; and when the meter is below the line, to the side of the tube.[40]

Dall Flow Tube

The Dall flow tube is a modified venturi tube that was developed in England during the 1940s. It has found wide use in gas, water, and steam flow applications and fluids which do not contain settleable solids. The basic design of this device is similar to the classical venturi; however, it is considerably shorter in length. Figure 3.9 illustrates the device.

The Dall tube has a lower permanent pressure loss than the classical venturi for a given flow rate and differential. It does, however, have a higher differential than a venturi with the same β ratio. Its differential pressure drop is nearly twice that of the venturi (at the same β and flow rate).

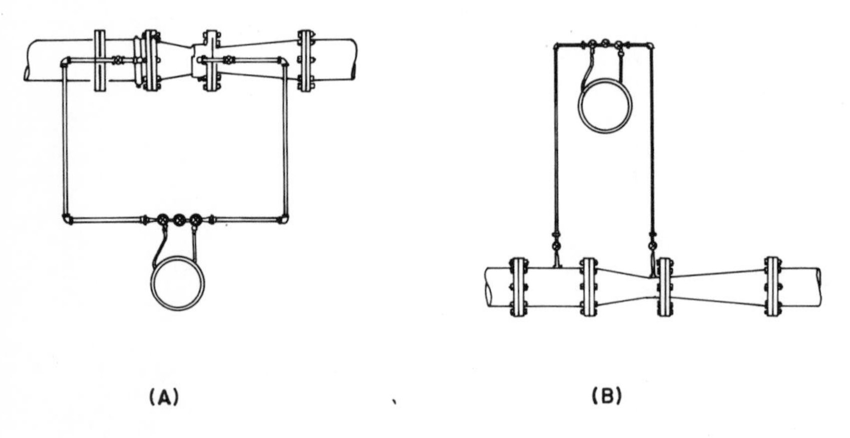

Figure 3.8 (A) Proper installation for liquid flow. (B) Proper installation for gas flow.

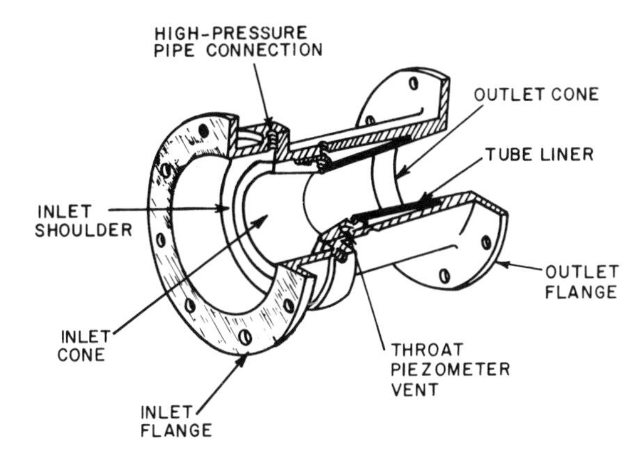

Figure 3.9 The Dall tube.

As shown in Figure 3.9, the device has a short, straight inlet section, the end of which decreases in diameter to the inlet shoulder. It has a converging cone section, a narrow annular gap or slotted throat annulus and a diverging outlet cone. The throat tap is located over the annulus, and the inlet-pressure tap is positioned directly upstream of the inlet shoulder.[40]

Venturi Accuracy and Operational Characteristics

In general, the venturi can supply extremely good accuracy in flow measurement. A properly calibrated unit can provide accurate measurements within ±0.5% for nearly all range of sizes. Venturis will maintain their accuracy over relatively long periods of time. The venturi is a self-cleaning device in most applications as its internal configuration allows smooth flow and efficient pressure recovery and minimizes erosion and clogging.

For the most part, venturi tubes are maintenance free. They have no moving parts nor mechanical features or glass that can undergo fatigue, strain, or breakage.

Probably the greatest advantage of a venturi is its low pressure loss as compared to other types of head meters. Figure 3.10 shows a comparison of the pressure loss in a venturi and Dall tube to other head meters. Pressure recovery is generally smooth and gradual within a minimum length of pipe after the fluid has passed through the throat area.[36]

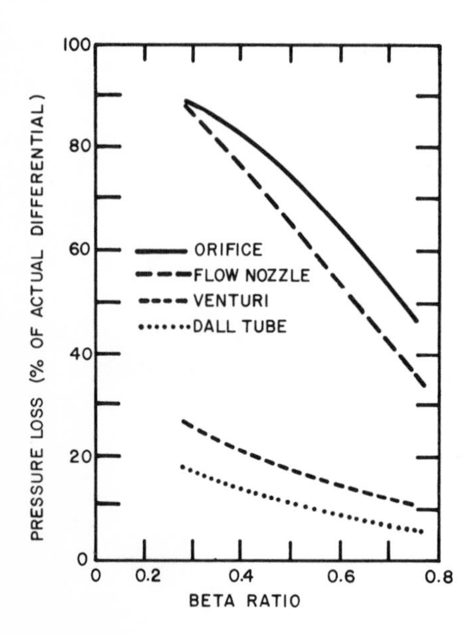

Figure 3.10 Comparison of the pressure loss in a venturi to other head devices.

3.3 THE ORIFICE PLATE

The orifice plate is one of the most widely used restrictions for differential pressure measurements in both liquid and gas applications. Its design is relatively simple and is generally low in cost and easy to install. Figure 3.11 shows the basic design of the orifice plate and the resultant flow patterns of a fluid in motion.

The differential pressure between the upstream and downstream sides of the unit are measured with pressure taps located on either side of the orifice plate. The most common design for the orifice is a circular hole in a metal diaphragm which is mounted concentrically between the flanges in the flow pipe. Other common designs are the eccentric and segmental types shown in Figure 3.12. The majority of orifice meters in use are of the concentric type. Orifices can be installed between the existing flanges in a piping system. On new installations, however, manufactured orifice flanged unions with built-in pressure taps are recommended.

Flange-mounted orifice plates offer the advantage of flexibility. They may be inserted and removed from a line without disturbing the piping or the differential pressure connections. Note that any increase in the flow rate of a fluid under pressure will cause a decrease in the pressure. The difference in pressures caused by changes in flow rates is a measure of the flow. With an orifice plate, a highly accurate prediction of the contraction of the flow stream (Figure 3.11) can be made.

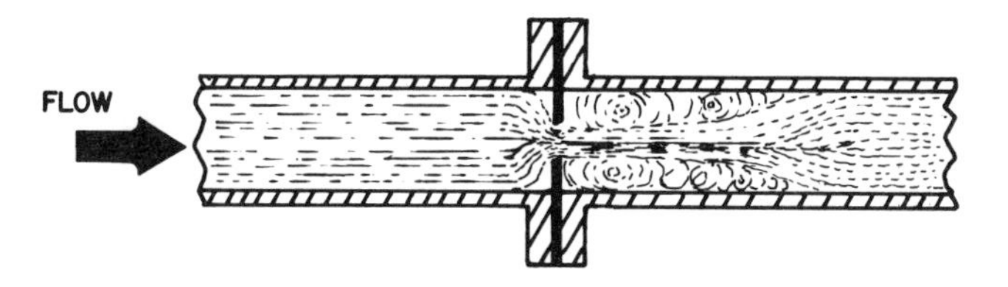

Figure 3.11 Basic design of the orifice plate and the flow patterns of a fluid passing through it.

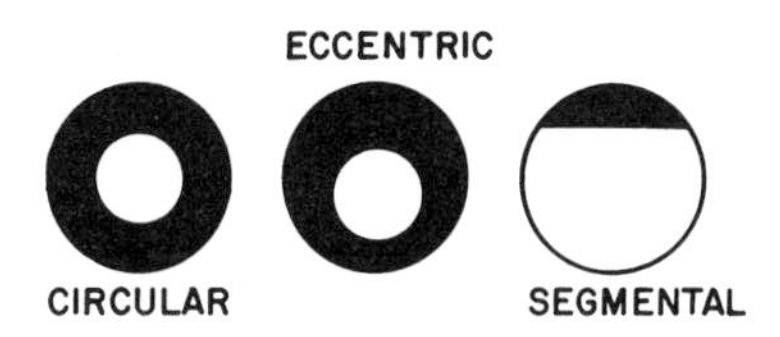

Figure 3.12 Various types of orifice shapes.

The orifice plate's upstream edge should be sharp and squared. The sharp edge should always be installed upstream as there are no appropriate correction factors if the beveled side faces upstream.

Orifice plates cannot be used for two-directional flow. There are no coefficients or operating characteristics that can be applied if the device is improperly installed.

An important design consideration is the position of the orifice opening, which should be accurately centered in the conduit. Improper location can again cause erroneous ΔP measurements.

Disadvantages of the sharp-edged, concentric design include inability to handle dirty fluids and viscous flows, inadequate disposal of condensate in flowing steam and vapors, and higher pressure loss than in the venturi (refer to Figure 3.10).

Eccentric Design

The eccentric design shown in Figure 3.12 is bored tangent to a circle concentric with the pipe and having a diameter roughly 98% that of the pipe. It is important that no portion of the hole be covered by the flange or gasket upon installation.

Segmental Orifice Design

This design consists of a segment of a circle located in an arc concentric with the pipe and with a diameter approximately 98% that of the pipe. The diameter ratio for this design is defined as the square root of the area ratio and is therefore a fictitious value.

The design has been used for metering wet steam flow, liquids containing particulates, and oils containing water.

For horizontal lines containing vapor or gas flows, the unit is used with the opening at the top of the pipe. It is not recommended for use with highly viscous fluids or liquids containing sticky solids as deposits build up on the edge of the orifice.

Pressure Taps

Pressure taps must be installed in such a manner as to prevent debris and sediment from clogging lines and to maintain the device sealed with the fluid. Furthermore, provision should be made for allowing gases or vapors to return to the line. Gas taps should always be positioned at the top of the unit or pipe. Measurement taps for steam and gas applications should be mounted in the side of the unit or pipe.

The most widely used taps for orifices in this country are of the flange type. They are positioned 25.4 mm from both the upstream and downstream faces of the plate.

The location of the upstream tap is somewhat arbitrary, except on high diameter ratios; however, the downstream tap location is critical. The most reliable measurements are obtained when the tap is located at the vena contracta, where the pressure profile is flat. For most orifice diameter ratios, the location can be roughly one-half pipe diameter downstream from the inlet face.

Downstream from the vena contracta a highly unstable flow regime exists, and pressure taps in this region should be avoided.[41]

Vena contracta taps are located one pipe diameter upstream and approximately one-half pipe diameter downstream. Usually these types of taps are not used on pipe diameters less than 10.2 cm.[40]

Taps are made through the pipe wall and flush with the inside pipe surface. For maximum pressure drops, pipe taps are positioned two and one-half pipe diameters upstream and eight diameters downstream.[40]

Incompressible Flow Equations

In a manner similar to the development of the expressions for venturis, semiempirical flow equations can be obtained for the orifice plate. For incompressible flow the actual volumetric flow can be written as follows:

$$Q_a = kA_2 \sqrt{\frac{2g_c}{\rho}(P_1 - P_2)} \tag{3.16}$$

where k is known as the flow coefficient given by

$$k = CM \tag{3.17}$$

Here C and M are defined as before.

Compressible Flow Equations

The expansion factor for an orifice employing flange taps or vena contracta taps is given by the following semiempirical expression:

$$Y = 1 - \left[0.41 + 0.35\left(\frac{A_2}{A_1}\right)^2\right]\frac{P_1 - P_2}{g\rho P_1} \tag{3.18}$$

(note A_2 is the area of the hole).

For orifices with pipe taps, the expansion factor is given by

$$Y = 1 - \left[0.333 + 1.145\left(\frac{A_2}{A_1} + 0.7\left(\frac{A_2}{A_1}\right)^{5/2} + 12\left(\frac{A_2}{A_1}\right)^{13/2}\right)\right]\frac{P_1 - P_2}{g\rho P_1} \tag{3.19}$$

or in terms of the ratio.

$$Y = 1 - [0.333 + 1.145(\beta^2 + 0.7\beta^5 + 12\beta^{13})]\frac{P_1 - P_2}{g\rho P_1} \tag{3.20}$$

And the actual mass flow rate is given as

$$\dot{m}_a = YkA_2\sqrt{2g_c\rho_1(P_1 - P_2)} \tag{3.21}$$

For convenience Table 3.2 tabulates values of the velocity of approach factor, M [Eq. (3.16)] and orifice ratios (i.e., β values; also applicable to venturi calculations).

The Discharge Coefficient

The flow coefficient, k, or more precisely the discharge coefficient, varies with fluid properties and flow conditions and is graphically related to the Reynolds number with the β ratio as a parameter. As an example, Figure 3.13 shows the variation of the flow coefficient, C, as a function of the throat Reynolds number for air flow through a venturi, fabricated in the laboratory from fiberglass-reinforced mat.[42]

This type of curve can be based on a Reynolds number definition at the throat conditions as shown (or based on upstream conditions if so desired).

$$Re_T = \frac{\rho u_m d}{\mu} \tag{3.22}$$

where Re_T is the throat Reynolds number and u_m the mean flow velocity. The mass flow is

$$\dot{m} = \rho u_m A_m \tag{3.23}$$

Here A_m is the cross-sectional area for the flow where u_m is measured.

Table 3.2
Tabulated Values of the Velocity-of-Approach Factor (M) Computer from Eq. (3.6)

β	M		M
1	1.0000	0.55	1.0492
0.05	1.0000	0.60	1.0719
0.10	1.0001	0.65	1.1033
0.15	1.0003	0.70	1.1472
0.20	1.0008	0.75	1.2095
0.25	1.0020	0.80	1.3014
0.30	1.0041	0.85	1.4464
0.35	1.0076	0.90	1.7052
0.40	1.0131	0.95	2.3219
0.45	1.0212	0.99	5.0377
0.50	1.0328		

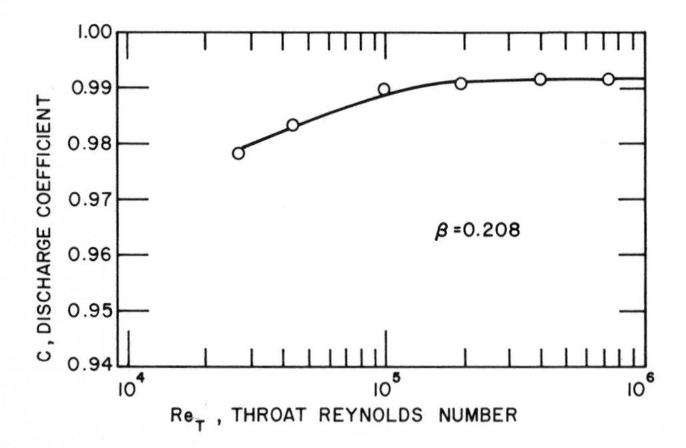

Figure 3.13 Illustrates how the discharge coefficient in a homemade venturi varies with the Reynolds number. (From Ref. 42.)

For throat Reynolds numbers above 200,000 the coefficient for a venturi tube can be assumed to be 0.99 to 1.00.

For an orifice plate, an approximation of the apparent flow can be made by assuming $C = 0.62$.

For more precise flow calculations, it will be necessary to account for fluid properties and flow conditions through the Reynolds number. Flow coefficients for orifice plates, venturis, and nozzles can be found in two ASME publications[37,38] and in other reference books, e.g., see Tuve.[43]

Example. Water is flowing at a rate of 325 GPH (gallons per hour) through a 10.2-cm (4-in.) i.d. pipe. If an orifice plate is installed in the line (discharge coefficient is 0.975), what is the pressure differential across corner taps? The orifice plate opening is 81% of the cross-sectional area of the pipe and the conditions are ambient.

Solution: From Eq. (3.12), $\beta = \sqrt{A_2/A_1} = d/D = \sqrt{0.81} = 0.90$. From Table 3.2, $M = 1.7052$. Convert volumetric flow rate to mass flow:

$$\dot{m}_a = \frac{(325)(0.1337)}{3600}(62.4) = 0.7532 \text{ lb/s}$$

Use either Eq. (3.16) or (3.21) (where $Y = 1$ for incompressible flow) to solve for P:

$$\dot{m}_a = kA_2\sqrt{2g_c\rho(P_1 - P_2)}$$

$$k = CM = 0.975 \times 1.7052 = 1.663$$

$$A_2 = \frac{1}{4}\pi\left(\frac{4}{12}\right)^2 = 8.73 \times 10^{-2} \text{ ft}^2$$

$$0.7532 = (1.663)(8.73 \times 10^{-2})\sqrt{2(62.4)(32.174)\,\Delta P}$$

$$\Delta P = 6.703 \times 10^{-3} \text{ lb}_f/\text{ft}^2$$

$$= 0.321 \text{ N/m}^2$$

Automatic Flow Rate Controller

Orifice plates have found wide acceptance in automatic flow control systems. One such design combines an adjustable orifice and an automatic internal regulating valve.

The regulating valve is positioned by a force balance of the differential pressure across the orifice. The force balancing action maintains constant differential pressure across the orifice, thus maintaining the flow rate as determined by the size of the orifice.[44]

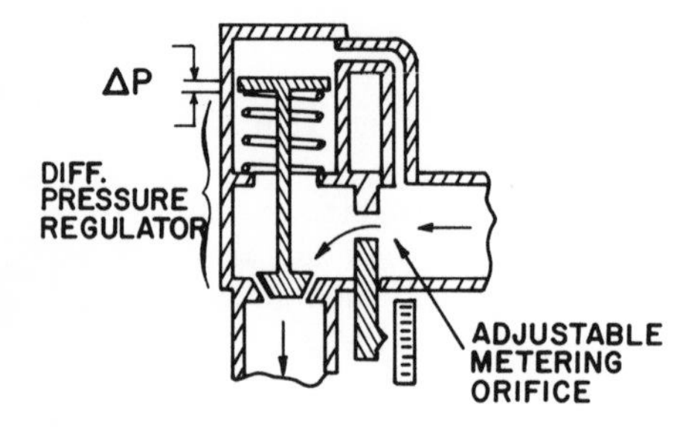

Figure 3.14 Illustrates the principle behind the automatic flow rate controller with adjustable orifice. (Courtesy of the W. A. Kates Co., Deerfield, Ill.)

The principle of operation is illustrated in Figure 3.14. The orifice is rectangular, narrow, long, and adjustable in length. This allows a uniform linear scale to indicate area and hence flow rate.

A manufacturer's specific design is illustrated in Figure 3.15. The orifice is an accurate slot in a vertical cylindrical sleeve. The sleeve encompasses a second cylinder, which is slotted similarly. The outer sleeve can be rotated about the inner one. When the two slots coincide, the metering orifice opening is maximum. When the slots are in another position, the measuring orifice area is directly proportional to the angle of rotation.[44]

The regulating valve is located in the orifice sleeve-cylinder mechanism. The valve includes a valve sleeve which is capable of sliding on a valve tube and can close or open valve ports. An impeller disk is driven downwards by the pressure differential across the orifice, causing the valve sleeve to close the ports. The force of a spring resists this closing action. The sleeve's position results from the direct balance of the two forces, and it sizes the valve port openings to cause the force balance.

The force balance maintains the orifice pressure differential constant regardless of upstream or downstream pressures or the size of the orifice opening. Complete corrective action has extremely fast response as control elements are direct-connected and directly actuated by the pressures of the fluid medium being controlled. Lag times are generally on the order of 1 to 2 seconds.

For remote stepless adjustment of controller flow rate set point, units can be fabricated with pneumatic or electric actuators. The actuator operation can be from a manual loading station or can be integrated into other process controls (e.g., temperature, level, velocity).

This type of design has been used in a variety of applications; some are illustrated in Figures 3.16 and 3.17. Figure 3.16A shows one application: a cascaded final control element of a complete pneumatic or electric control system employing a throttling actuator. The set point is changed by a system

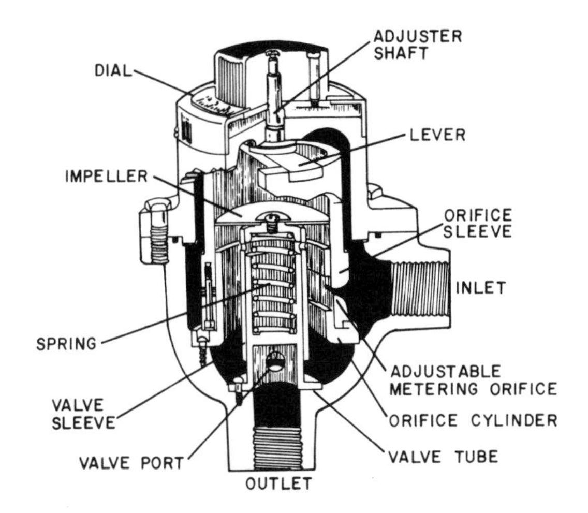

Figure 3.15 Illustrates one manufacturer's design of the automatic flow rate controller. (Courtesy of the W. A. Kates Co., Deerfield, Ill.)

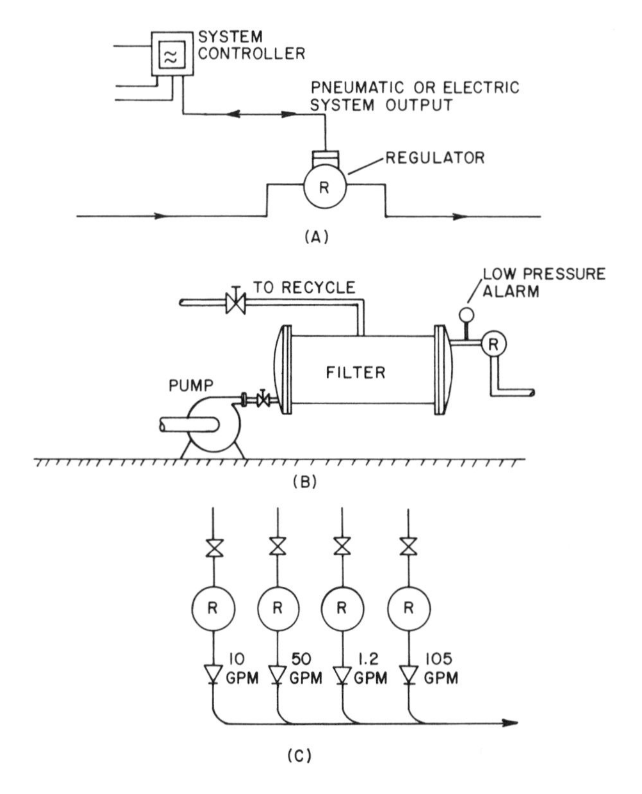

Figure 3.16 Various applications of the automatic flow rate controller. (A) Illustrates a cascaded final control element. (B) Application to pressure filtration. (C) Application to proportionate blending. (Courtesy of the W. A. Kates Co., Deerfield, Ill.)

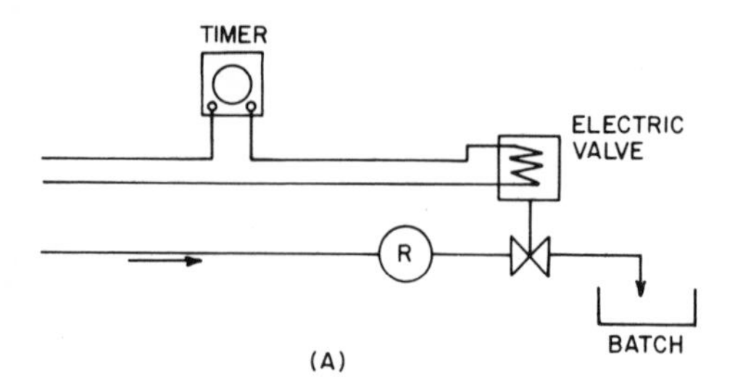

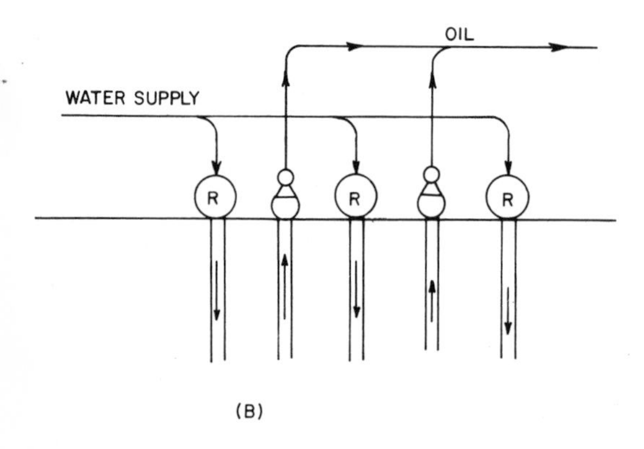

Figure 3.17 Various applications of the automatic flow rate controller. (A) Application to a batch-operated process. (B) Application to a water-flooding oil recovery operation. (Courtesy of the W. A. Kates Co., Deerfield, Ill.)

controller. In Figure 3.16B the unit is applied to a pressure filter where the drop across the filter increases as the filter cake builds up. A controller in the outlet line maintains the filter effluent constant at the allowable or safe rate. The controller can also be reset to suit downstream process requirements as warranted. As shown in the figure, a low pressure alarm system can be used to signify the end of a filtration run.

Figure 3.16C illustrates another application: the blending of several liquids. Continuous proportionate blending of liquids can be achieved despite large, erratic variations in line pressures by the use of a controller in each line. An operator can preset each controller for the desired flow rates.

Figure 3.17A illustrates the metering of liquids in a batch operated process. The scheme shown employs a controller, a timer, and an automatic stop valve. Figure 3.17B illustrates a water-flooding oil recovery operation. Controllers maintain the drive rate for any injection station, regardless of the water supply pressure or formation backpressure.[44]

Other typical applications include corn processing, oil production, the pulp and paper industry, sewage treatment, the pharmaceutical industry, water treatment, the plastics industry, and various other industrial chemical processes.

3.4 FLOW NOZZLES

One of the first literature citations of the commercial use of flow nozzles dates back to the mid-nineteenth century.[45] In 1934, the ASME initiated an extensive research program on flow nozzle coefficients. Several of these studies are summarized in Refs. 46-48.

The basic design of the standard flow nozzle is illustrated in Figure 3.18. As shown, the rounded approach has a curvature that is equivalent to the quadrant of an ellipse. The distance between the front face and tip is roughly one-half the pipe diameter (where the throat ratio of the nozzle is between 0.4 and 0.8). For long radius nozzles, throat ratios are usually less than 0.4. The discharge coefficient for a flow nozzle generally lies between that of an orifice plate and a venturi. For a rough approximation of the apparent flow, the discharge coefficient can be assumed to have a value of 0.98. The effect of Reynolds number on flow nozzle discharge coefficient is affected primarily by the type finish of the inlet and throat regions. Rough finishes produce a flatter coefficient over a wider range of Reynolds numbers. Special calibration of the device is however, required.

The incompressible and compressible flow equations presented for orifice plates are directly applicable to nozzles as well. For more accurate calculations, detailed computations and values of discharge coefficients for flow nozzles can be found in Refs. 37 and 38.

Applications

Flow nozzles have found wide use in the measurement of wet gases (e.g., saturated steam with condensate in suspension). Entrainment in gas

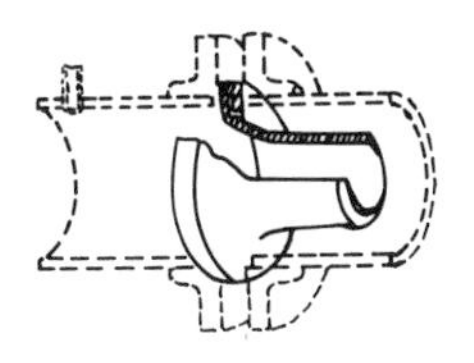

Figure 3.18 Basic design of the flow nozzle.

or vapor streams can cause excessive erosion problems on some types of head meters; however, the nozzle's curved surface face guards tend to minimize such action.

Flow nozzles have been used in the metering of high-velocity fluids. This application is most often used when plant capacities undergo an increase with no changes in piping. In the production of high-temperature steam, for example, pipe sizes are usually maintained at a minimum due to the high accelerating costs of high-temperature piping as pipe diameter increases. Flow nozzles are used in these cases to meter resultant high-velocity high-temperature steam flow.[49]

Characteristics

Flow nozzles are more efficient than orifice plates. In general, they can handle roughly 60% more fluid than an orifice plate at the same pressure drop.[40] Another advantage over orifice plates is that nozzles are capable of handling liquids with suspended solids. When metering liquids with suspended solids, it is generally recommended that the flow should be vertically downward. This allows some suspended matter to drop out of suspension and will limit variations in the approach conditions as there will be no place for matter to lodge. The streamlined design of nozzles causes solids to be swept through the throat.

Flow nozzles are not suited for highly viscous fluids or fluids containing large amounts of sticky solids (generally magnetic flowmeters are preferred for the latter, if the fluid is conductive).

Table 3.3 provides a rough guide to the range of operating characteristics of flow nozzles and other head meters discussed thus far.

Sonic Nozzles

Sonic flow nozzles have been the subject of considerable investigation for a number of years.[50] Sonic flow conditions can exist at minimum flow area when the flow rate and pressure differential become sufficiently high. The flow is described as being "choked," and flow rates achieve maximum values at inlet conditions.[51] For isentropic flow and assuming an ideal gas, the pressure ratio at the choked or critical condition can be expressed as follows:

$$\left(\frac{P_2}{P_1}\right)_{crit} = \left(\frac{2}{\gamma + 1}\right)^{\gamma/(\gamma-1)} \tag{3.24}$$

where $\gamma = C_P/C_V$.

An expression for the mass flow rate can be obtained by substituting Eq. (3.24) into the mass flow expression given by Eq. (3.9):

$$\dot{m} = A_2 P_1 \sqrt{\frac{2g_c}{RT_1}} \left[\frac{\gamma}{\gamma + 1} \left(\frac{2}{\gamma + 1}\right)^{2/(\gamma-1)}\right]^{1/2} \tag{3.25}$$

Table 3.3
Characteristics of Flowmeters[a]

Meter	Range of max. flow (GPM)	Max. pressure (psig)	Temp. range (° F)	Max. viscosity (cP)	Construction materials	Tolerance (%)
Orifice	0.2–3500	6000	-455 to 2000	4000	Most metals	1-2
Flow nozzle	0.5–15,000	1500	-60 to 1500	4000	Bronze, iron, steel	>2
Venturi	0.5–1500	1500	-60 to 1500	4000	Bronze, iron, steel, plastics	>1

[a]From Cheremisinoff and Niles.[39]

Equation (3.25) allows calculation of the ideal sonic-nozzle mass flow rate. To compute the actual flow rate, a discharge coefficient should be included in the expression. A value of 0.97 for C is applicable for most engineering calculations. The criterion for Eq. (3.25) is that only upstream stagnation conditions can be used for values of P_1 and T_1.

3.5 PITOT TUBES

The pitot tube is one of the oldest basic devices used for measuring velocity and has found applications in measurement of both liquid and gas flows. There are a variety of designs. Two of these, shown in Figure 3.19, are of the single-tip type that incorporates static-pressure connections. In general, they consist of two concentric tubes; the inner tube transmits the impact pressure while the annular space between the two tubes transmits static pressure. The static pressure taps shown are located well back from the tube's front face to prevent interaction from eddy currents. Common designs are generally of the sharp- or round-nose fashions.

Manometers are generally used as the pressure-indicating device, as in the case of the venturi meter. The range of rates that pitot tubes are capable of measuring (using standard secondary differential measuring devices) is narrow and hence has limited their industrial usage. In addition, the ratio of average-to-center velocities in pipes varies through wide limits; thus, it is necessary to preform a velocity traverse over the pipe diameter to obtain a meaningful average velocity. Figure 3.20 shows the data obtained from a 10-point traverse over a 63.5-mm I.D. pipe for air flow. The plot shown is in terms of the square root of the impact head. Readings obtained from a pitot tube are a direct measure of the velocity head ΔH.

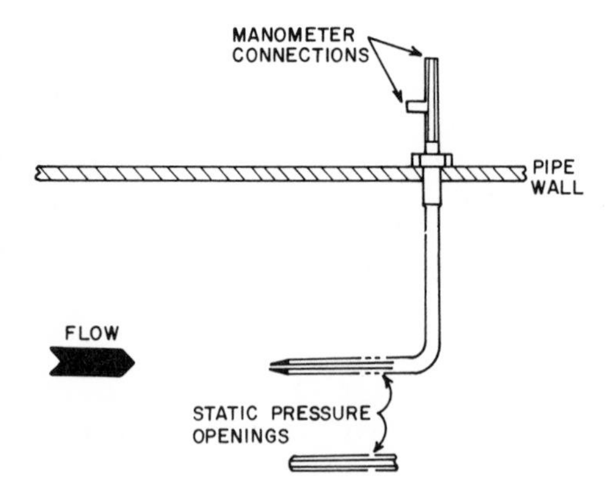

Figure 3.19 L-type pitot tube designs.

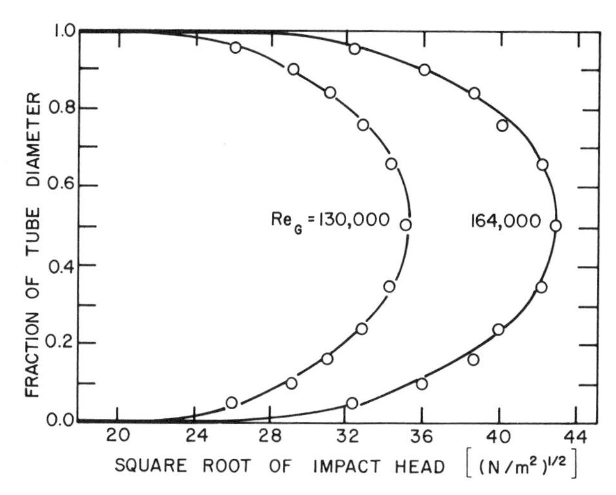

Figure 3.20 Typical velocity head profiles for a vertical traverse across a 63.5-mm i.d. pipe with air flow. (From Ref. 42.)

For ΔH expressed in millimeters of fluid flowing, the velocity at any point along the diameter is given by

$$u_0 = C\sqrt{2g_c\ \Delta H} = C\sqrt{2g_c(P_1 - P_0)/\rho_0} \tag{3.26}$$

A proper traverse is made by dividing the cross-sectional area of flow into a number of equal subareas and taking a local measurement at a representative point within each subarea. Numerically integrating over the measured velocity profile gives the mean velocity (integration can be done by Simpson's rule). The greater the number of subdivisions of flow area, the greater precision achieved in the average or mean tube velocity. For a 20-point traverse, the average velocity is good to within ±0.1%. For a rough approximation (good to within ±5%) a single measurement at the center of the tube times 0.9 is sufficient.

If quantity rate measurement is needed, it can be computed from the ratio of average velocity to the velocity at the point of measurement. Figure 3.21 shows the ratio of average-to-center velocity for air flow through a smooth acrylic pipe. At high Reynolds numbers, pipe roughness causes an increase in the relative center velocity.

Disadvantages

The primary disadvantage of the pitot tube is that it is limited to only point measurements (although this is an advantage in some applications). In addition, pitot tubes are very susceptible to fouling by foreign matter in the

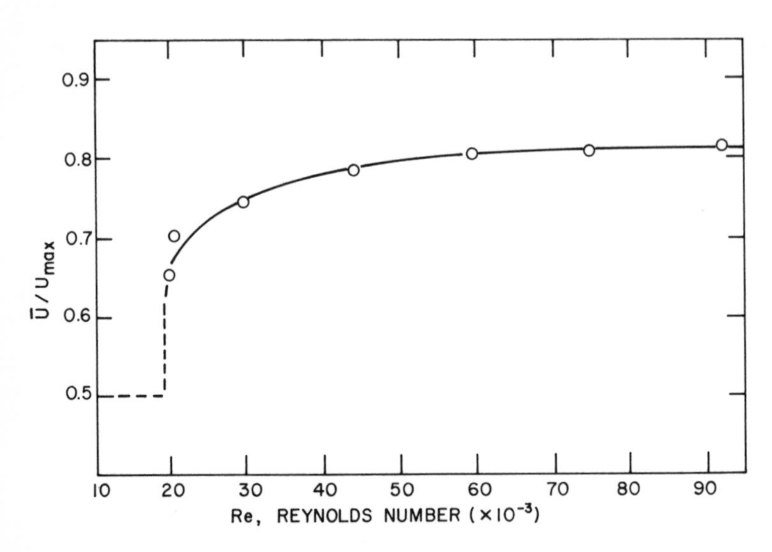

Figure 3.21 Ratio of average velocity ($\bar{u}$) to center or maximum velocity (u_{max}) plotted as a function of the Reynolds number (where Re is based on $\bar{u}$). Data obtained for air flow in a 63.5-mm i.d. smooth pipe. (From Ref. 42.)

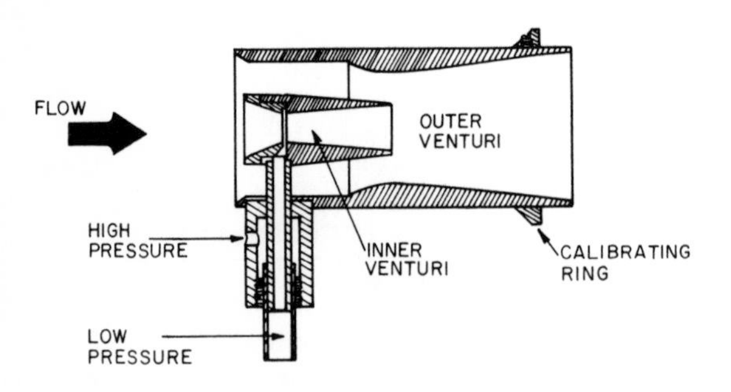

Figure 3.22 Pitot-venturi design.

fluid medium. Another shortcoming is that the pressure drops which they create are often too small to be detected accurately by standard differential pressure meters.

Pitot-Venturi Design

In order to overcome the differential pressure effects, the pitot-venturi design, shown in Figure 3.22, has evolved. Single- or double-venturi sections may be added to a pitot tube to increase the pressure differential.

The venturi head essentially decreases the static pressure, thus creating a magnified differential. No measurable pressure loss results from the use of this design.

NOMENCLATURE

A	Area (m^2 or ft^2)
C	Discharge coefficient defined by Eq. (3.4)
C_A	Correction factor defined in Eq. (3.14)
C_P	Specific heat at constant pressure ($Btu/lb_m \cdot °F$ or $cal/g \cdot °C$)
C_S	Correction factor defined in Eq. (3.15)
C_V	Specific heat at constant volume ($Btu/lb_m \cdot °F$ or $cal/g \cdot °C$)
D	Diameter (m or ft)
d	Throat diameter (m or ft)
F_{pa}	Pressure correction factor for air [refer to Eq. (3.14)]
F_{ps}	Pressure correction factor for steam flow [refer to Eq. (3.15)]
F_{ta}	Temperature correction factor for air [refer to Eq. (3.14)]
g_c	Conversion factor, 32.174 $lb_m \cdot ft/lb_f \cdot s^2$ or 4.17×10^8 $lb_m \cdot ft/lb_f \cdot hr^2$
ΔH	Velocity head (mm)
k	Flow coefficient [refer to Eq. (3.17)]
M	Velocity-of-approach factor
$\dot{m}$	Mass flow rate (kg/s or lb_m/s)
P	Pressure (N/m^2 or psi)
ΔP	Pressure differential (N/m^2 or psi)
Q	Volumetric flow rate (m^3/s or ft^3/s)
Q_f	Volumetric flow defined in Eq. (3.13)
R	Universal gas constant (1544 $ft \cdot lb_f/lb \cdot mol \cdot °R$, 1.986 $Btu/lb \cdot mol \cdot °R$, 1.986 $cal/g \cdot mol \cdot K$)
Re_G	Gas phase Reynolds number
Re_T	Throat Reynolds number, Eq. (3.22)
S_g	Specific gravity
T	Temperature (K or °R)
u	Velocity (m/s or ft/s)
u_m	Mean flow velocity at throat (m/s or ft/s)
Y	Expansion factor

Greek Letters

β	Ratio of throat to tube diameter [refer to Eq. (3.12)]
γ	C_P/C_V
ρ	Density (g/liter or lb_m/ft^3)

Subscripts

a	Refers to actual conditions
crit	Refers to critical

i Refers to ideal or theoretical
0 Refers to local conditions
1 Refers to upstream conditions
2 Refers to downstream or throat conditions

PROBLEMS

3.1 A venturi tube is to be used to measure the flow rate of water through a 3-in. i.d. line. The maximum flow that can be expected is 140 GPM.
(a) Compute the upstream Reynolds number.
(b) If the pressure differential across the meter at maximum flow conditions is measured to be 9 psi, compute the dimensions of the venturi.
(c) Compute the throat Reynolds number. The density and viscosity of water are 62.4 lb_m/ft^3 and 2.36 $lb_m/h \cdot ft$, respectively.

3.2 Kerosene is flowing through a 12-in. venturi with a 7-in. throat. The differential pressure is measured to be 72 in. of water. Compute the volumetric and mass flow rates. (The viscosity of kerosene is 1.85 cP at 95° F, and its specific gravity is approximately 0.8.)

3.3 Air is flowing through a 10-in. high-pressure line at line pressure 352.8 psig. A venturi section has a throat diameter of 6-in. and the pressure drop is measured to be 150 in. of water. How many pounds per hour of air is being passed? (At 0° C and 25 psi, $C_P/C_V \simeq 1.47$.)

3.4 Methane (specific gravity with respect to air = 0.55) is flowing through an orifice plate with β ratio 0.70. The line pressure is 190 psig, and i.d. is 8 in. The orifice plate uses pipe taps, and the differential pressure is measured to be 180 in. of water. Compute the mass flow rate.

3.5 For the orifice plate in Problem 3.4, using flange taps compute the flow rate.

3.6 Air flow through a 4-in. line is being measured with a sonic nozzle. Pressure and temperature measurements taken upstream are 735 psia and 170° F, respectively. (Note: $C_P/C_V \simeq 1.53$.) The mass flow rate is 4.5 lb_m/s.
(a) Compute the critical pressure ratio.
(b) Compute the throat diameter, assuming critical flow conditions.

3.7 In Problem 3.6, upstream pressure and temperature measurements are assumed to be at stagnation conditions. If the pressure was obtained as a static-pressure measurement, justify the solution to the problem. (Note: $C_P \simeq 0.241$ Btu/$lb_m \cdot$ °F.)

4

Positive Displacement Flowmeters

4.1 INTRODUCTION

Positive displacement type flowmeters are commonly employed where a consistently high degree of accuracy is required under steady flow conditions. These devices are accurate with well-defined tolerances, normally ±1% over flow ranges of up to 20:1. They display relatively low pressure drops and are well suited to batch operations and mixing or blending systems.

Positive displacement meters function by channeling portions of the flow into separate volumes, according to the meter's physical dimensions, and measures the flow by counting or totalizing these volumes.

Positive displacement devices are further characterized by one or more moving parts, positioned in the flow stream. These components physically separate the fluid into volumetric increments. Mechanical parts are driven by energy supplied from the fluid motion, which produces a pressure loss between the inlet and outlet of the device. The accuracy of these systems largely depends on clearances between moving and stationary components. As such, these meters require precision-machined parts for the small clearances necessary. In general, this class of flowmeters is not adaptable to metering slurries or fluids with appreciable amounts of suspended particles.

The positive displacement meters described in this chapter include reciprocating piston, nutating disk, rotary piston, and rotary vane meters.

4.2 APPLICATIONS

Positive displacement flowmeters have been used in a variety of liquid handling operations ranging from the household water meter to a spectrum of industrial applications. Several of these industrial applications are briefly described in this section.

Industrial Chemicals. Positive displacement meters and accessories are fabricated from materials to match almost any chemical or physical compatibility requirement. Manufacturers claim they can fabricate this type of meter to measure virtually every type of liquid.

Pharmaceuticals. These systems can be applied to metering in any operation in this industry ranging from barrel filling to batching to continuous blending, with relatively high accuracy.

Paints and Varnishes. Meters have been used to accurately control input quantity of liquid components in batch processes for the production of paints and varnishes. Push-button presets and high meter performance helps to assure speed, accuracy, and dependability.

Cryogenics. These meters accurately meter liquids at cryogenic temperatures. Low temperatures generally do not affect seals or measurement accuracy.

Pipeline Applications. Temperature-compensated meters have been employed with advanced accessories, in capacities in excess of 2000 GPM. Specific applications have been to crude and refined products covering a wide range of viscosities.

Oil Field Production. Meters equipped with automatic temperature compensators have been used for accurate metering of petroleum from the wellhead.

Liquefied Gases. Meters have been used for the retail or wholesale distribution of liquefied petroleum gas (LPG), butane, anhydrous ammonia, and carbon dioxide.

Liquid Fertilizers. Meters are available in a range of construction materials for corrosion resistance to almost all farm chemicals and fertilizers. Material selection includes stainless steel and aluminum.

Fuel Oil Delivery. Meters are available in virtually all capacities for this application. Units generally are designed for minimum field maintenance.

Table 4.1 gives examples of specific liquids handled by this class of meters.

4.3 RECIPROCATING PISTON METERS

Reciprocating piston meters are of the single and multipiston types. The specific choice of design depends on the range of flow rates experienced in

Table 4.1
Examples of Specific Liquids Handled by Positive Displacement Meters

Diethanolamine	Acetates
Phosphoric acid	Cyclohexanone
Methylene chloride	Liquefied petroleum gas (LPG)
Gasoline	Glycols
Liquid nitrogen	Honey
Toluene	Latex
Vinegar	Fuel oil
Water	Liquid oxygen
Syrup	Caustics
Asphalt	Liquid sugar
Xylene	Butadiene
Fruit juices	Acetone
Crude oil	Liquid lard
Cyclohexylamine	Whiskey
Jet fuel	Fish oil
Alcohols	Ammonia
Styrene	Acrylonitrile
Cresylic acid	Formalin

an application. With a suitable choice of construction materials, the reciprocating piston meter can be employed to meter practically all liquids within ±1 accuracy. (Extreme accuracy is achievable to 0.1% for these designs.) Periodic lubrication is required.

The piston meter is essentially a piston pump operating backwards. Figure 4.1 illustrates the operating principle of the reciprocating single-piston meter. The piston undergoes back and forth motion from the fluid under pressure. As the piston moves, reaching the end of its stroke, it shifts the intake and discharge valve. This in turn operates a counter which adds a fixed volume increment of fluid with each cycle or stroke of the piston. Figure 4.2 illustrates a typical multipiston design.

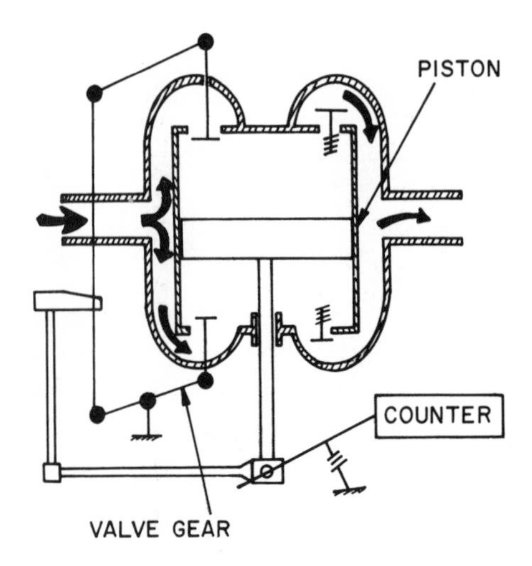

Figure 4.1 Operating principle of the reciprocating piston meter.

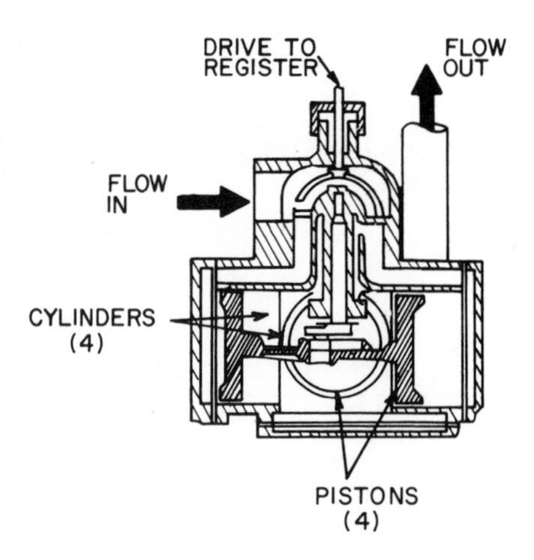

Figure 4.2 Reciprocating multipiston meter design.

4.4 NUTATING DISK METERS

The nutating disk meter consists of a movable disk mounted on a concentric sphere. The disk is contained in a working chamber with spherical sidewalls and top and bottom surfaces that extend conically inward. It is restricted from rotating about its own axis by a radial partition which extends across the entire height of the working chamber. The disk is slotted to fit over this partition. Figure 4.3 shows a cutaway view of the meter.

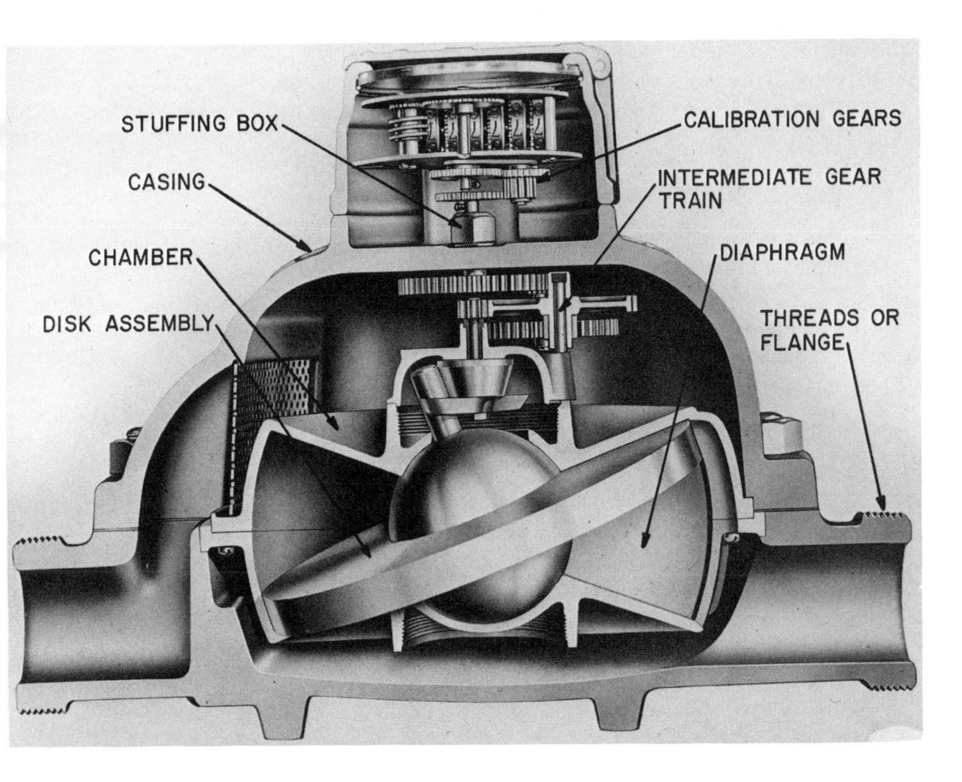

Figure 4.3 Cutaway view of a nutating disk meter. (Courtesy of Hersey Products Inc., Spartanburg, S.C.)

In the figure, the liquid enters the left side of the meter and strikes the disk, forcing it to rock (nutate) in a circular path without rotating about its own axis. A pin extending out from the inner sphere, perpendicular to the disk, traces a circular path as it is driven by the nutating motion. This pin drives the undergear which controls the meter's register.[39]

It is important to note that the nutating disk is the only moving part in the measuring chamber. There are several disk materials available for use with a broad range of liquids: measuring chambers are available in bronze, iron and steel; Ni-Resist and stainless steel chambers are available in a range of sizes.[51,52] Bronze meters are employed for use with most liquids that are noncorrosive to bronze, for example, water, glycol, brine, liquid sugar, and fish solubles. Iron meters are used with most liquids that are noncorrosive to iron. Examples are water, animal fats, fuel oils, vegetable oils, molasses, benzene, gasoline, and coal tar distillates.

Stainless steel meters can service most liquids, such as nitric acid, fruit juices, deionized water, phenol, formaldehyde, and oleum. The direct-drive mechanical shaft couples the meter to the register through a stuffing box (refer to Figure 4.3). Models are also available with magnetic drive.

Factors Affecting Meter Accuracy

The accuracy rating of nutating disk liquid meters is normally expressed as a percentage variation over the full recommended flow range (note: not as a percent of the full flow value). This means that the percentage variation is improved as the actual operating flow range is reduced.

Figure 4.4 shows the performance curve for one manufacturer's meter. By changing the ratio of the calibration gears, the meter's performance curve can be vertically shifted so that the midpoint of the actual flow range is 100% accurate. (This, of course, largely depends on the fluid being handled.)

The performance curves for all types of displacement meters will change form, with changes in fluid viscosity. Variations in viscosity affect the meter's slip characteristic, which in turn causes variations in flow accuracy. The slip factor in these types of meters are a result of the clearances. Fixed manufacturing tolerances for the measuring chamber and nutating disk provide the clearances between moving and stationary components.

Manufacturing clearances are usually scaled to match a specific viscosity range. Additional accuracy adjustment is provided by the meter calibration gears.

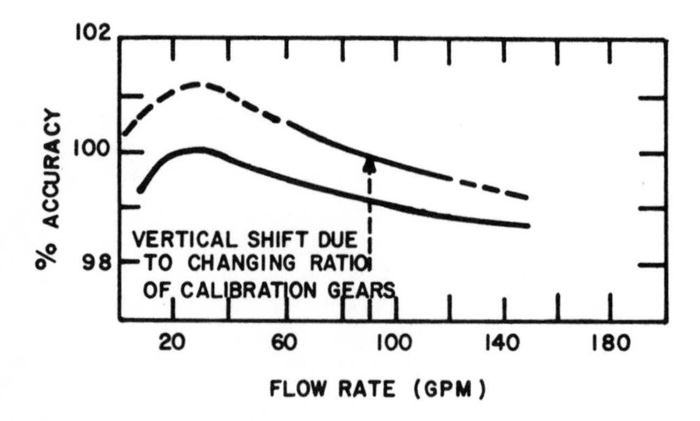

Figure 4.4 Performance curve of a nutating disk meter. The bottom curve represents the accuracy over the total flow range. By changing the accuracy of the calibration gears, the accuracy curve shifts upward. (Courtesy of Hersey Products Inc., Spartanburg, S.C.)

Increases in viscosity causes a proportional increase in the pressure drop (see Chapter 2). As such, the recommended flow range is reduced with an increase in viscosity in order to maintain a realistic service life factor. Table 4.2 shows the accuracy variation of one manufacturer's nutating disk meter over various viscosity ranges.

Temperature can also play a key role in meter accuracy. A significant increase in fluid temperature will result in predictable changes in the dimensions of the internal meter parts. Meter parts are machined to operate within a certain temperature range. For example, a meter that has been built to operate in a temperature range of 350-500° F will have excessive clearances and consequently proportionate inaccuracy when operated at 90° F. Conversely, a meter that is specified for a temperature operating range of 50-80° F cannot provide satisfactory service at 250° F.

Meter Selection

A general guide to selecting the optimum meter for a specific fluid flow application is given below:

1. Specify the materials of construction.
2. Verify temperature range or limitation for operating conditions.
3. Verify the operating pressure.
4. Specify the meter size.
5. Select the meter register most suitable to the particular application.

A brief discussion of each step follows.

Materials of Construction

The selection of meter materials should be given careful consideration, especially when the metering corrosive liquid is extended. Chemical suppliers should be consulted for specific recommendations on materials selection. Tables 4.3 and 4.4 can be used as a rough guide to material selection for nutating disk meters. The viscosity groups given in Table 4.3 are explained in Table 4.5.

Table 4.2
Variation of Meter Accuracy with Viscosity[a]

Viscosity (SSU)	Variation (%)	Viscosity (SSU)	Variation (%)
30-75	1.0	125-150	1.0
75-100	1.0	150-200	1.5
100-125	1.0	200-10,000	~3.0

[a]From Ref. 52.

Table 4.3
Rough Guide to Meter Material Selection[a,b]

Liquid[c]	Meter construction	Disk/ball material[d]	Viscosity group[e]
* Acetaldehyde	Iron	M	1
‡* Acetic acid	Stainless steel	M	2
* Acetic anhydride	Stainless steel	D	2
Acetone	Bronze	C	1
Alcohol (ethyl or methyl)	Bronze	A	1
Alcohol (denatured)	Bronze or iron	C	1
* Alum solution	Stainless steel	M	2
* Aluminum nitrate 5%	Stainless steel	M	< 2
* Aluminum sulfate 50%	Stainless steel	M	<2
* Ammonia (anhydrous cold)	Iron or steel	M	2
* Ammonia (aqueous)	Iron	A or M	2
* Ammonium hydroxide	Iron	A or M	2
‡* Ammonium nitrate	Iron or stainless steel	A or M	2
* Ammonium phosphate	Stainless steel	M	2
Amyl acetate	Bronze	C	2
* Analine	Iron (black)	M	2
‡* Animal fat	Iron	D	3
* Apple juice	Stainless steel	H or M	2
Asphalt (mastic)	Iron	D	6
‡* Benzine	Iron	D	1
Benzol	Bronze	D	1
* Black liquor soap	Stainless steel	M	2
Brine (sodium)	Bronze	A	2
Bunker C oil	Iron	D	5
Butadiene	Iron	C	1
Butadyne	Bronze	C	1

Table 4.3
(continued)

	Liquid[c]	Meter construction	Disk/ball material[d]	Viscosity group[e]
	Buttermilk (for cattle feed)	Bronze or iron	A or H	2
*	Butylamine	Iron	C	3
*	Calcium chloride 30%	Iron	A or H	<2
*	Carbon bisulfide	Iron	M	1
	Carbon tetrachloride	Iron	C	1
	Casein	Iron	H	2
‡*	Caustic soda	Iron or stainless steel	A or M	2
*	Chloroform	Iron	C	1
*	Chocolate liquor	Iron (black)	D	2
*	Chrome liquor	Stainless steel	M	2
*	Citrus fruit juices	Stainless steel	M	2
*	Coal tar distillate	Iron	D	6
*	Cocoa butter	Iron (black)	D	4
*	Coconut oil	Iron	D	3
	Coffee (hot)	Bronze	C	1
	Core oil	Iron	D	3
‡*	Corn oil	Iron	D	3
*	Cottonseed oil	Iron (black)	D	3
	Corn syrup	Iron	D	6
	Creosote	Bronze or iron	D	4
	Cutting oil	Bronze or iron	H	4
*	Cyanide solution	Iron	A	2
	DDT solution	Iron	H	1
*	Dibutyl phthalate	Iron	M	2
*	Diethylamine	Iron	C	1
*	Distilled water	Stainless steel	C or M	1

Table 4.3
(continued)

Liquid[c]	Meter construction	Disk/ball material[d]	Viscosity group[e]
Emulsion oil & water	Bronze or iron	C or M	5
Ester (aromatic)	Bronze	C or D	2
* Ether	Iron	C	1
* Ether (ethyl)	Steel or stainless steel	C	1
Ethyl acetate	Iron	C	1
* Ethylene diamine	Stainless steel	C	3
Ethylene glycol	Iron	D	2
* Ferric sulfate solution	Stainless steel	A	1
* Formaldehyde	Stainless steel	M	1
Fuel oils Nos. 1 & 2	Iron	D	2
Nos. 3 & 4	Iron	D	3
Nos. 5 & 6	Iron	D	4
Fish oil	Iron	D	3
Fish solubles	Bronze or iron	C	2
Freon	Bronze or iron	C	1
Gasoline	Bronze or iron	D	1
Glue	Iron	D	6
Glycerine	Bronze or iron	D	2
Grease	Iron	D	6
* Hydrogen peroxide	Stainless steel	F	2
Isobutylene	Iron	C	1
* Isopropylamine	Iron	D	1
Kerosene	Bronze or iron	D	1
Ketones	Iron	C	1
Lacquer	Iron	D	3
* Lactic acid	Stainless steel	M	2

Table 4.3
(continued)

	Liquid[c]	Meter construction	Disk/ball material[d]	Viscosity group[e]
‡*	Lard (molten)	Iron (black)	D	3
	Latex solution	Iron	D or H	6
	Lecithin	Iron	A	1
*	Lime sulfur solution	Iron	M	1
*	Liquid soap solution	Iron	H	2
	Malt syrup	Bronze	H	3
*	Methyl or ethyl acrylate	Iron	D	1
*	Methyl ethyl ketone	Iron	C	1
*	Methyl formate	Steel	C	1
	Mineral oil	Iron	D	2
	Mineral spirits	Iron	D	1
	Molasses (cold)	Iron	H	6
	(hot)	Iron	D or H	5
*	Monochlorobenzene	Iron (black)	C	2
	Monochlorobenzol	Bronze	D	1
	Naphtha	Iron	D	1
*	Naphthenic acids	Stainless steel	C	5
*	Nitrogen solutions	Iron or stainless steel	A or M	2
*	Nitric acid	Stainless steel	F	2
	Oil (soluble cutting)	Iron	H	2
‡*	Oleic acid (red oil)	Iron	D	2
*	Oleum	Stainless steel	F	2
*	Organic acid	Stainless steel	M	2
	Paracol wax	Iron	D	3
	Paraffin (molten)	Iron	I	3
	Paint (oil base)	Iron	D	3

Table 4.3
(continued)

Liquid[c]	Meter construction	Disk/ball material[d]	Viscosity group[e]
Pentachlorophenol	Bronze	C	3
Perchloroethylene	Iron	C	2
* Phenol	Stainless steel	M	2
Phenolic resin	Iron	D	6
* Pineapple juice	Stainless steel	M	2
Plasticizer	Iron	C or D	3
Printing ink	Iron	D	6
* Polyvinyl chloride resin	Stainless steel	M	2
* Potassium chloride 30% (cold)	Iron	A, H, M	<2
* Potassium aluminum sulfate	Iron	A or M	2
* Potassium hydroxide 50%	Iron or steel	M	<2
Resin emulsion	Iron	H	4
Resin polyester	Iron	C	3
Resin size	Bronze	H	3
Rubber cement	Iron	H	6
‡* Salad oil	Iron (black)	D	3
‡* Soap	Iron	M	2
Soap (resin)	Iron	A or H	3
* Soda ash	Iron	H or M	2
‡* Sodium carbonate	Iron	A or M	2
‡* Sodium hydroxide	Iron or steel	C or M	2
Sodium silicate	Bronze or iron	A or D	6
‡* Soya oil	Iron	D	3
Soy bean oil	Iron	D	3
* Stearic acid (fatty acid)	Stainless steel	D or M	3

Table 4.3
(continued)

	Liquid[c]	Meter construction	Disk/ball material[d]	Viscosity group[e]
	Stoddart solvent	Iron	D	1
	Sugar cane juice	Bronze	I	2
	Sugar (liquid)	Bronze or iron	H	2
*	Sulfide liquor	Stainless steel	M	2
*	Sulfuric acid 94%	Iron or stainless steel	F	>2
*	Tall oil	Stainless steel	C	3
‡*	Tallow (molten)	Iron	D	3
*	Tanning liquor	Iron	M	2
	Tar	Iron	D	5
	Tetrachloroethane	Iron	D	1
	Thinners	Bronze or iron	D	1
	Toluene	Bronze or iron	D	1
	Toxaphene & DDT	Iron	C	2
	Trichlorethylene	Iron	C	1
	Turpentine	Iron	D	2
	Vanilla extract	Iron	A	2
	Varsol	Iron	D	1
‡*	Vegetable fat or oil	Iron	D	2
‡*	Vinegar	Stainless steel	A or M	2
	Vinsol	Iron	C	1
*	Vinyl acetate	Stainless steel	M	1
‡*	Vinyl chloride	Iron	D	2
	Vinyl plastics paint	Iron	D	4
*	Viscose	Iron or stainless steel	M	4
	Water, cold to 100°F	Bronze or iron	A	2

Table 4.3
(continued)

Liquid[c]	Meter construction	Disk/ball material[d]	Viscosity group[e]
Water, 100-180°F	Bronze or iron	A	2
above 180°F	Bronze or iron	C or H	4
* Water deionized, cold	Stainless steel	M	2
hot	Stainless steel	M	1
Water gas tar	Bronze or iron	D	4
Wax emulsion	Iron	H	2
Wax (hot)	Iron	D	1
Weed killer	Bronze or iron	C	2
Whey	Bronze or iron	A or H	2
Whiskey	Stainless steel	F	2
* Wine (cold)	Stainless steel	A or H	2
Xylol (xylene)	Iron	C or D	1

[a]Courtesy of Hersey Products Inc., Spartanburg, S.C.
[b]Use in conjunction with Tables 4.4 and 4.5.
[c]Key: * Indicates stainless steel chamber and gear train should be used.
‡ Indicates Ni-Resist chamber and gear train may be used.

Temperature Limitations

Verifying operating temperatures will help to establish a range of viscosities, meter construction materials, and degree of flow accuracy that can be achieved by the meter.

Operating Pressure

All meters will have a maximum pressure rating at a specific temperature or over a range of temperatures. In general, steel and bronze meters can be used for relatively high pressure applications.

Meter Size

Manufacturers will supply tables and/or charts that will provide information on the meter's flow capacity range from which a specific size meter

Table 4.4
Material Selection Guide for Disk and Ball[a]

Disk and ball	Letter code	Maximum temperature rating
Hard rubber	A	100° F
Carbon	C	400° F
Aluminum	D	400° F[b]
Kel-F	F	200° F

Disk/ball	Letter code	Maximum temperature rating
Spauldite/carbon	H	225° F
Aluminum/carbon	I	Same as aluminum
Kel-F/carbon	M	200° F

[a]Courtesy of Hersey Products Inc., Spartanburg, S.C.
[b]For liquids between 200° F and 300° F, disk diameter is as follows:
undersized 0.010 in. on meter sizes 0.75 through 2.50 in.
undersized 0.020 in. on meter sizes 3 and 4 in.
undersized 0.030 in. on meter size 6 in.
For higher temperatures consult the manufacturer.

can be selected to meet the flow requirements. Meter capacities are normally tabulated along with viscosity groups (as in Table 4.5) and meter size.

In specifying meter size, the anticipated pressure drop through the meter must be determined in order to confirm the meter's compatibility with flow conditions required. Figure 4.5 is a nomograph for estimating the pressure drop across a meter. The nomograph is based on tests performed by one manufacturer on their stock nutating disk meters; however, it is useful for rough estimates. The example at the end of this section illustrates the use of Figure 4.5.

Meter Register

There are a variety of meter registers available, each suited for a particular application. Registers fall under the following general classifications according to application groupings:

Table 4.5
Viscosity Groups in Table 4.3[a]

	Viscosity range		
Group	SSU	cP	Examples
1	30	0.20-0.75	Freon 12, dipropyl ketone, ether, toluene, MEK (methyl ethyl ketone), carbon tetrachloride
2	31-450	1-90	Water (<80° F), olive oil, ammonia, blood, sulfuric acid, turpentine, SEA 10 oil (80° F)
3	450-1000	90-220	SAE 30 oil (100° F), Grade 4 road tar (120° F), sucrose 64 to 68 Brix (70° F)
4	1000-5000	220-1100	Paints, castor oil, ketchup, glycerine (70° F), black liquor (122° F)
5	5000-20,000	1100-4400	Table molasses, wood resin (200° F), bunker C oil (100° F)
6	20,000-50,000	4400-11,000	Honey (70° F), blackstrap molasses (70° F), Grade 8 road tar (>100° F), corn syrup 41° B at 80° F

[a]From Ref. 52.

1. Totalizing registers
2. Flow rate indicators
3. Manual batch registers
4. Electricontact batch registers
5. Batch controllers

Figures 4.6 and 4.7 show a totalizing register and one type of batch controller, respectively.

Example. A nutating disk meter is to be used in a 1-in. nominal size line. If the maximum pressure drop across the meter is not to exceed 23 psi, what is the minimum flow rate that can be expected? Assume water flow.

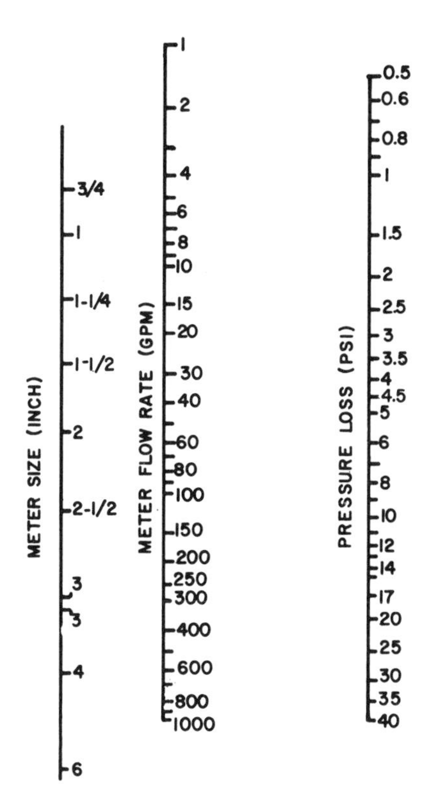

Figure 4.5 Nomograph for determining the pressure loss across a nutating disk meter. (Courtesy of Hersey Products Inc., Spartanburg, S.C.)

Solution:
Step 1. Locate the meter size value on the first column of the nomograph.
Step 2. Locate the pressure loss value on the right-hand column of the nomograph.
Step 3. Draw a straight line through the two points, and read the flow rate (29 GPM) off of the middle column. (Figure 4.8 illustrates this.)

Note that the nomograph is for water flow only. Pressure losses or flow rates obtained must be corrected for liquids with different viscosities.

Figure 4.6 A mechanical totalizing register which provides continuous digital totalized readings. The unit is nonresettable. The chemical resistant register incorporates self-lubricating gears for extended service. The register is housed in a high-impact, chemical-resistant polymer register box with thick glass lens. This unit is available with reed switch pulse output. (Courtesy of Hersey Products Inc., Spartanburg, S.C.)

Figure 4.7 A mechanical batch controller. The unit consists of a mechanical batch register and globe valve. These units can be mounted on nutating disk meters and provide automatic batch control with two-stage shut-off. The unit shown is equipped with emergency interrupt, five-digit preset, and eight-digit nonreset totalizer. It is available with explosion-proof switch for warning lights and pump control. (Courtesy of Hersey Products Inc., Spartanburg, S.C.)

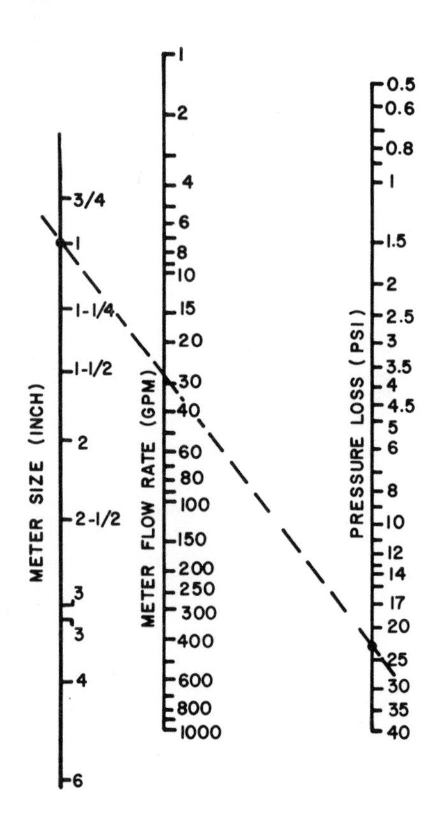

Figure 4.8 Illustrates use of the nomograph. To obtain pressure drop, locate meter size value and flow rate on left and middle columns, respectively. Draw a straight line through the two points, intersecting the right-hand column and read ΔP in psi.

4.5 ROTATING MECHANISM DESIGNS

Rotary Piston Meters

The basic design of the internal workings of a rotary piston meter is illustrated in Figure 4.9. A cylindrical working chamber houses a hollow cylindrical piston of equal length. The central hub of the piston is guided in a circular motion by two short inner cylinders. The piston and cylinder are alternately filled and emptied by the fluid flowing through the meter. A slot in the sidewall of the piston is removed such that a partition extending inward from the bore of the working chamber can be inserted. This restricts the piston to a sliding motion along the partition.

Rotary Vane Meters

Figure 4.10 illustrates a rotating vane type rotary meter. To ensure that the vanes are in continual contact with the meter casing, they are spring loaded. As the eccentric drum rotates, a fixed quantity of liquid is swept through each section to the outlet. The volume of the displaced fluid is measured by a register attached to the shaft of the eccentric drum. In general, these are highly reliable meters and relatively insensitive to fluid viscosity changes.

The rotating vane type rotary meter is a lightweight unit that can be easily installed in the field. Flow accuracy is established by machined castings that form the measurement compartments. These meters are employed in gas and liquid applications for distribution measurement in commercial and industrial areas and in-plant metering.

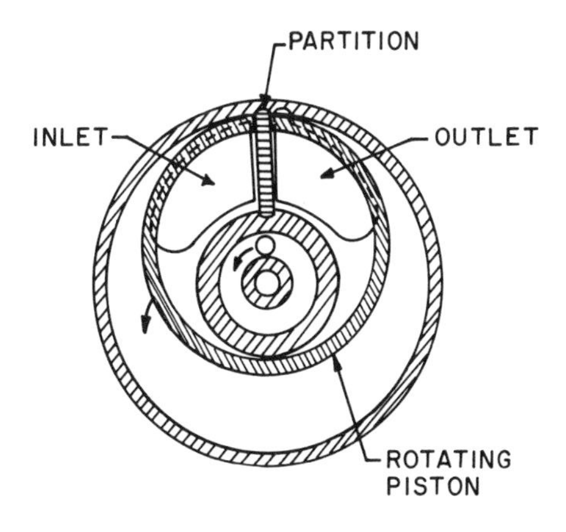

Figure 4.9 Internals of a rotary piston meter.

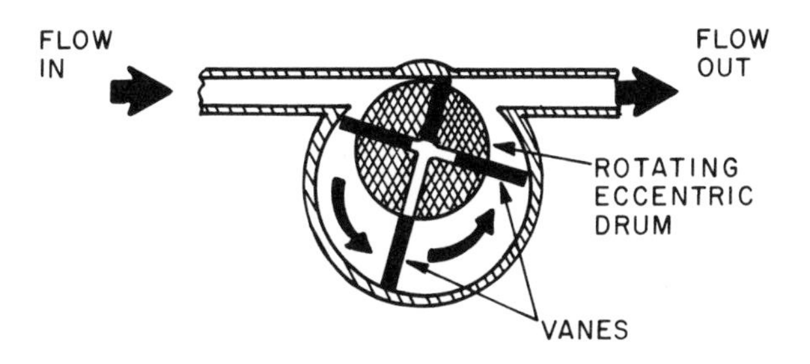

Figure 4.10 Illustrates the basic design of a rotary vane flowmeter.

For gas applications, it is generally good practice to purge the gas line prior to installation of the meter. In lines that contain high amounts of pipe scale or other particulates, the use filters is recommended to extend meter service life.

4.6 OTHER POSITIVE DISPLACEMENT FLOWMETERS

An additional differential pressure flowmeter worth noting is a calibrated flow nozzle design. The flow through a calibrated nozzle is sensed by an arrangement of opposed bellows. Displacement of the bellows is transferred by a low-friction cam and lever to a rotary geared movement to indicate flow rate. The operation of this design is entirely mechanical.[53]

Designs are self-contained and do not require separate orifices, blocking, purging, or equalizing valves. This type of flowmeter is well suited for applications where compactness, low cost, and shock or impact resistance are priority factors. Typical applications include metering of fuel oil to burners, purging of instrument lines, injection of chemicals into process streams, and monitoring flow of inert shielding gases.

5

Mechanical Flowmeters

5.1 INTRODUCTION

The rotameter and turbine meter are discussed in this chapter. Both devices accomplish flow measurement by drag effects. The drag coefficient is a function of the Reynolds number and hence depends on the fluid viscosity. As such, both instruments must be recalibrated for each new fluid handled, with calibrations periodically checked.

5.2 THE ROTAMETER

Rotameters are often referred to as variable-area meters. They are widely employed both in laboratory and industrial applications. The basic design, illustrated in Figure 5.1, consists of a tapered vertical tube in which the fluid enters through the bottom, causing a float or bob to ascend. The tube is narrow at the bottom, where the fluid velocity is the greatest, and widest at the top, where the velocity is at a minimum.

The float, used to indicate the flow rate, is slightly heavier than the fluid being measured, and as such will sink to the bottom of the tapered tube when there is no flow. As fluid flows through the tube, the float continues to rise until it reaches a point where the drag forces are balanced by the weight and buoyancy forces. The float's position in the tube, measured by a linear scale, is taken as an indication of the flow rate.

A force balance about the float in Figure 5.1 gives the following equation:

$$F_D + \rho_f \phi \frac{g}{g_c} = \rho_B \phi \frac{g}{g_c} \tag{5.1}$$

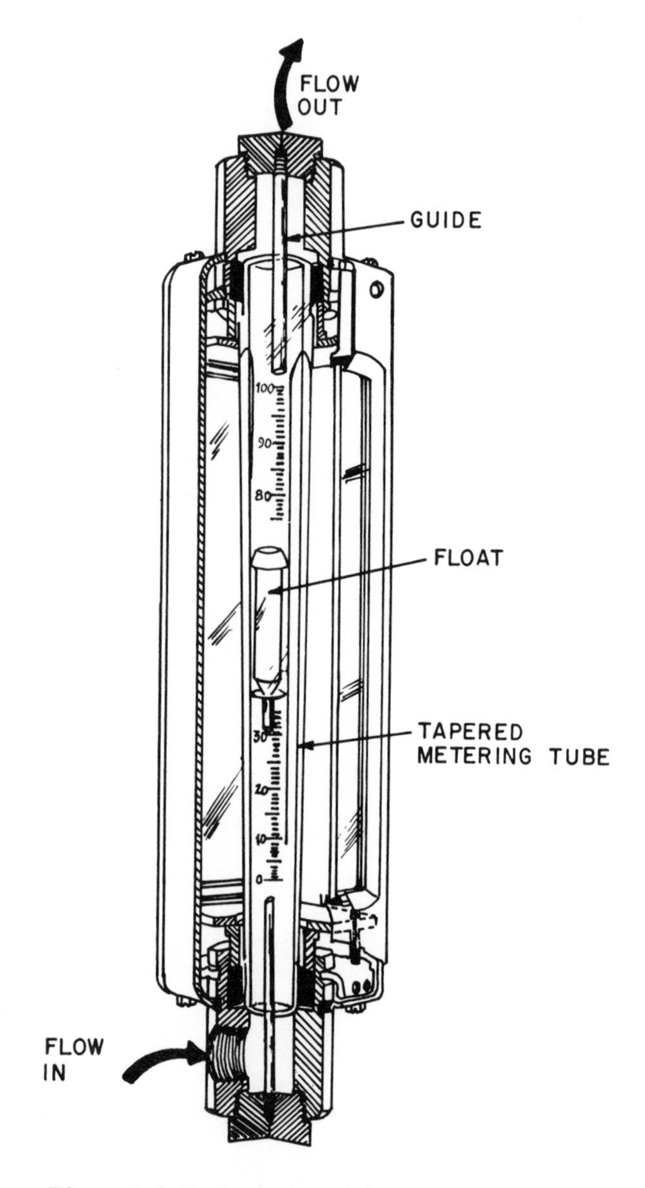

Figure 5.1 Basic design of the rotameter.

where ϕ represents the volume of the float, ρ the density, F_D the drag force, g the acceleration of gravity (32.2 ft/s^2), g_c a conversion factor, and subscripts f and B indicate fluid and bob (float), respectively.

The drag force, F_D, is defined by Eq. (5.2),

$$F_D = C_D A_B \frac{\rho_f U_m^2}{2g_c} \tag{5.2}$$

where

C_D = drag coefficient

A_B = area of the front face of the float

U_m = mean fluid flow velocity in the annular space between the float and tube wall

Substituting Eq. (5.2) into (5.1) and rearranging, an expression for the mean fluid velocity is obtained.

$$U_m = \left[\frac{2g\phi}{C_D A_B}\left(\frac{\rho_B - \rho_f}{\rho_f}\right)\right]^{1/2} \tag{5.3}$$

From this expression the volumetric flow rate can be obtained by multiplying U_m by the annular flow area ($Q = U_m A$), where

$$A = \frac{\pi}{4}\left[(D + \alpha z)^2 - d^2\right] \tag{5.4}$$

Here

D = diameter of the tube at the inlet (cm)

d = maximum bob diameter (cm)

z = vertical distance from the entrance (cm)

α = constant related to the tube taper

Approximating Mass Flow Rate

For approximations, and indeed for many practical meters, the area relationship given by Eq. (5.2) can be assumed linear for the actual tube and float dimensions. This is useful in that an expression for mass flow rate can be written simply as

$$\dot{m} = C'z\left[(\rho_B - \rho_f)\rho_f\right]^{1/2} \tag{5.5}$$

where C' is an appropriate meter constant obtained through calibration.

It is advantageous to employ a flowmeter that indicates measurements which are independent of fluid density variations. Special float construction can allow rotameters to be used under fluid density changes. In such cases the mass flow rate is expressed by

$$\dot{m} = \frac{C' z \rho_B}{2} \tag{5.6}$$

where the float's density is approximated by

$$\rho_B = 2\rho_f \tag{5.7}$$

Limitations and Operating Problems

In general, rotameters measure small flow rates relative to the flow capacities of other methods already discussed.

An operational problem frequently encountered with rotameters is that the float has a tendency to stick to the tube wall upon contact. In unstable flows, the float is often pushed against the tube wall. This problem is normally eliminated by the use of a guide wire which passes up through the center of the float and tube. Still another solution to this problem involves the use of glass ribs on the inside of the tube.

Rotameters are primarily limited to relatively clear fluids although there are methods that facilitate measurement of dark or opaque fluids. For example, rib walls can be fabricated into the tube to guide the float so as to allow such a small clearance that the float will be visible through the film between it and the tube wall. This particular design has the disadvantage of blockage or plugging from suspended particles in the fluid.

Another approach to handling dark fluids is to employ a transparent disk that is illuminated from the back wall of the tube, creating a line of light which allows reading the measurement.

Unstable or pulsating flow will cause the float to undergo vertical oscillations. The mean measurement of these oscillations does not necessarily represent the true flow through the rotameter. Only under ideal conditions where the oscillations of the float correspond to the amplitude of the flow pulsation does the average measurement provide the true flow.

Materials of Construction

Rotameters can be fabricated from a variety of materials. There is a broad line of acrylic plastic rotameters useful for measuring low flow rates of both liquids and gases. The meter body for these is usually supported in an extruded aluminum frame which can accommodate process connectors and absorb connection strain.

A metal-tube rotameter is illustrated in Figure 5.2. Steel or other metal can be used to house the float. The float's position can be detected by observing an indicating magnet which moves up and down with a mating magnet located within the metal extension tube.

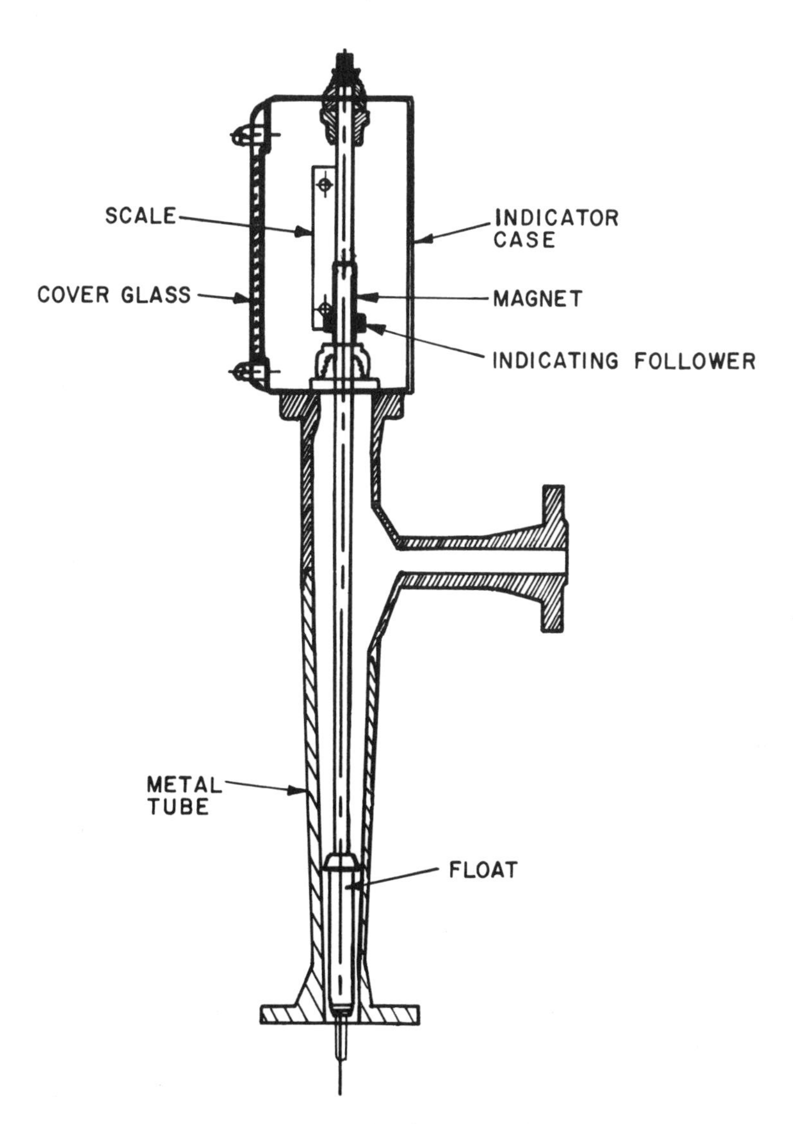

Figure 5.2 Illustrates a metal-tube rotameter.

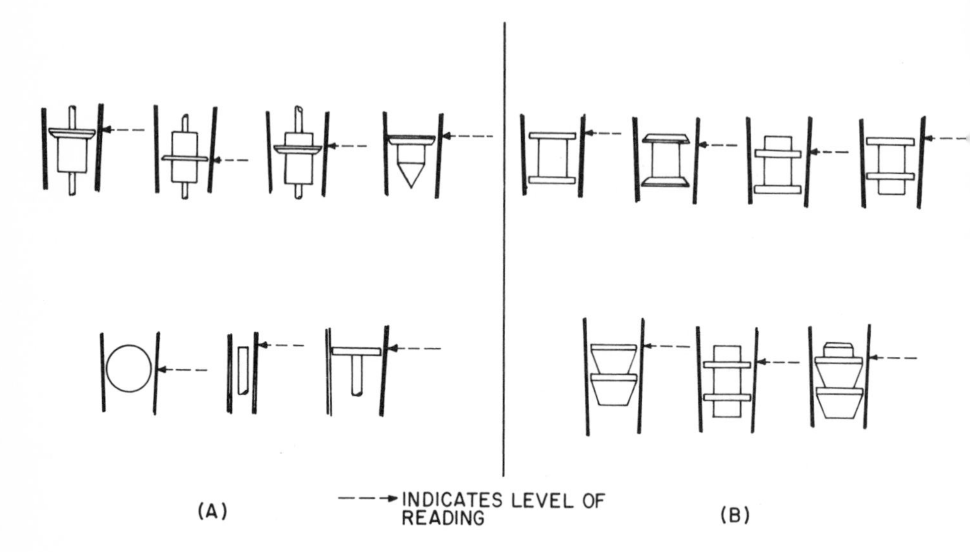

Figure 5.3 Illustrates various types of rotameter floats. (A) Extension-type floats. (B) Spool-type floats.

Types of Floats

There are a large number of float designs from which to choose. Two main classifications of floats are the extension type and spool type. Figure 5.3 illustrates several designs and indicates the proper manner in which they are read.

5.3 THE TURBINE METER

In principle, the turbine meter operates similarly to an hydraulic turbine. The turbine is positioned within a designed passage through which the fluid is directed. The fluid's impact on the blade surfaces causes rotation. The rotor can be positioned such that it can be driven by radial or axial flow, or a combination of both. The rotor motion may directly drive a register. Frequently, magnets on the rotor are employed to generate a rotating field and thus produce a current indicative of the flow rate. Figure 5.4 illustrates the basic design of the turbine meter.

In the figure, the fluid to be metered enters the flowmeter body chamber at the inlet port. The rotor assembly contained in the chamber consists of the turbine and hover plate elements mounted at each end of the shaft. The rotor assembly rests on the top hover seat in a no-flow condition. When flow begins, a pressure drop occurs across both hover plates and

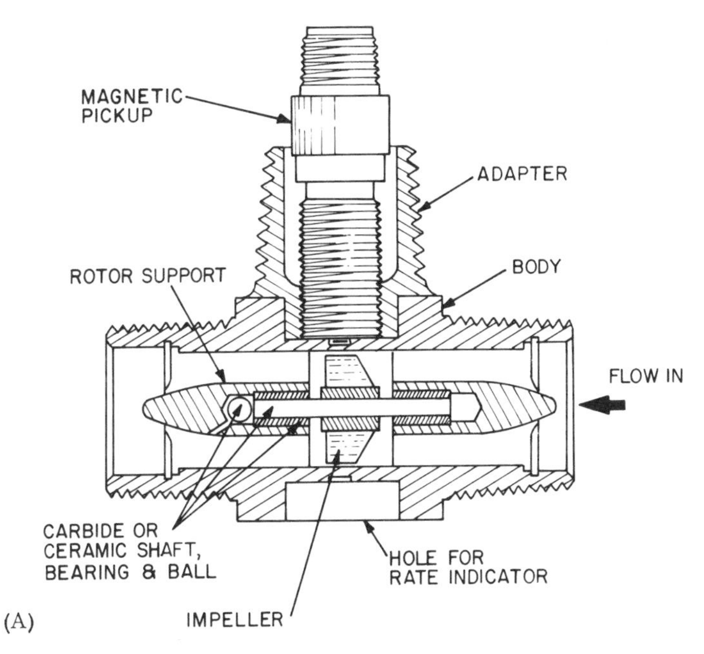

(A)

(B)

Figure 5.4 The turbine meter. (A) Illustrates the major components of the turbine meter. (B) Cutaway view of a turbine meter showing the internal workings of the model. (Courtesy of C-E Invalco, Combustion Engineering Co., Tulsa, Okla.)

corresponding seats. This pressure drop lifts the rotor assembly from the top seat. The fluid element is split within the body chamber, and the flow through the two turbines causes the rotor assembly to rotate with stability at a speed that is directly proportional to the flow velocity. The fluid leaves the meter body chamber through two tapered seating ports that lead the flow to the exit port.[54]

A magnetic pickup is employed (see Figure 5.4) to sense the rotational speed of the rotor, and an ac sinusoidal signal is generated. The frequency of these signals is directly proportional to the fluid flow rate. The ac signal is transmitted by means of a conductor-shielded cable to an appropriate readout instrument.

Applications

Turbine meters provide an accurate and relatively economical method of metering to a wide variety of liquids. In the food industry, for example, turbine meters have been employed in metering milk, cheese, whey, cream, syrups, vegetable oils, vinegar, etc.[55] Figure 5.5 shows one type of

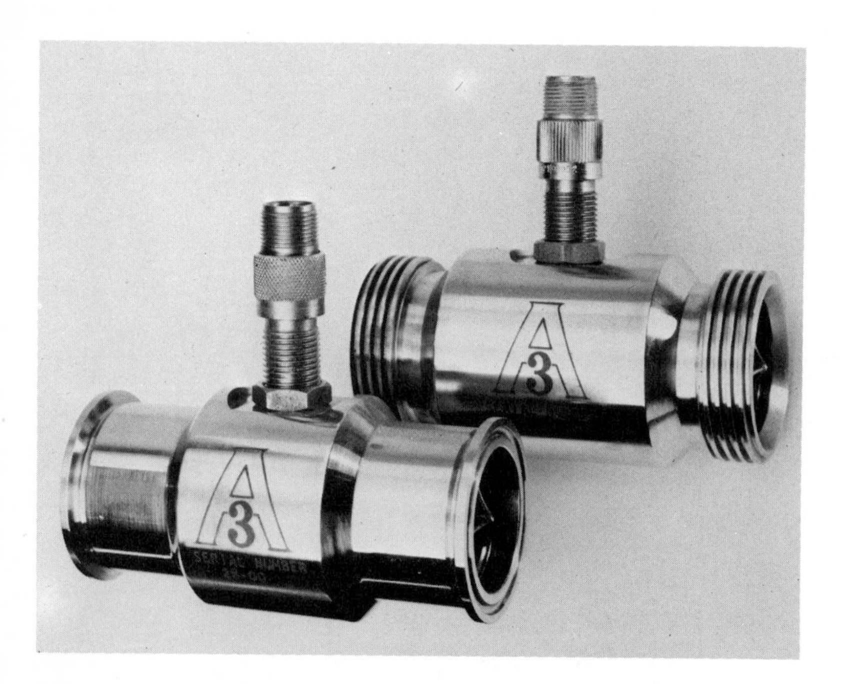

Figure 5.5 Typical turbine flowmeter employed in the beverage and food industries. (Courtesy of C-E Invalco, Combustion Engineering Co., Tulsa, Okla.)

sanitary turbine flowmeter widely used throughout the beverage and food processing industries.

Another application where turbine meters have found widespread use is in oil field automation. During the 1960s, oil companies began automating their production facilities employing computers for data storage of production statistics.[56] Computer utilization expedited data retrieval for oil, water production reports, gas produced, as well as oil and gas sold and water injected for water-flooding or pressure-maintenance purposes. Turbine flowmeters have been incorporated into these liquid flow systems. The wide range and variety of output signals has allowed these flowmeters to be interfaced with almost all data acquisition instrumentation.

In Figure 5.6 the turbine flowmeter is employed in an oil/water separator operation. These flowmeters can be used on two-phase and three-phase high- and low-pressure separators. Flowmeters are normally installed roughly ten pipe diameters of straight pipe upstream and five pipe diameters downstream. Valves are usually of the snap-acting type so that a constant flow rate can be provided each time the separator dumps. Pilot valves can be lever operated or actuated by liquid head provided that the pilot valves are pressure balanced against the static pressure head of the vessel.[56]

Another oil production application is in water injection projects. Oil that is produced from formation pressure (primary production) results in the recovery of roughly 30-35% crude oil. When the crude is pumped out with nothing to replace it, surface subsidence often results. This surface subsidence can be raised by means of injecting water back into the formation to replace the removed oil. The water injection method is also used to force more crude oil out of a given formation. The injected water usually consists of a produced water plus a certain quantity of makeup. Produced water is obtained from the crude oil because wells tend to capture salt water with the crude. Most injection water used is salt water, which presents a serious corrosion problem. Turbine flowmeters are different from other differential flow measuring devices in that they are capable of handling liquids that are corrosive or abrasive, as well as liquids containing appreciable solids and dissolved gases. Turbine flowmeters for metering injected water are tailored specifically to handle problem liquids.

There are a multitude of designs suited to particular liquid handling applications. Figure 5.7 shows still another turbine flowmeter design with flanged connections (see also Table 5.1). Table 5.2 is a partial listing of various liquids measured with turbine meters.

Readout Instrumentation. There are two major components in any flow metering system: the meter and the readout instrument. In the case of the turbine meter, the unit generates electrical pulses that are proportional to the flow. The readout instrument accepts these pulses and displays its reading directly in units such as gallons or barrels, in the form of totalized flow on a counter and/or flow rate on an indicating meter.

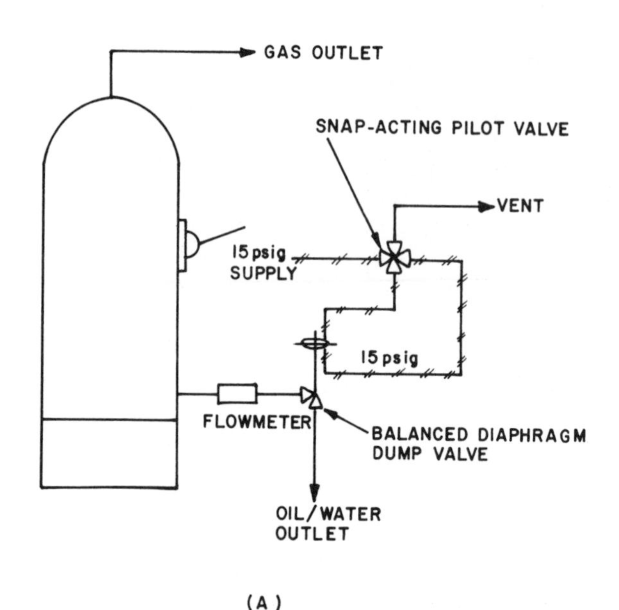

Figure 5.6 Turbine meter application to oil production. (A) Application to a two-phase separator. (B) Application to a three-phase separator. (Courtesy of Flow Technology, Inc., Phoenix, Ariz.)

Figure 5.7 Turbine flowmeter constructed from tungsten carbide sleeve bearings with flanged connections. Table 5.1 gives the flow range capability and frequency output for this particular model. (Courtesy of C-E Invalco, Combustion Engineering Co., Tulsa, Okla.)

Readout instruments for turbine meters fall under two basic categories: totalizers and rate indicators. Rate indicators are basically frequency-to-dc converters. They accept the turbine meter output frequency and convert the signal to dc to drive an electrical meter. The electrical meter can then be calibrated in any usable units [e.g., barrels per day (BPD) or gallons per minute (GPM)]. Figure 5.8 shows a basic instrumentation hookup for a flow rate indicator system.

The totalizer system (illustrated in Figure 5.9) accepts the pulses from the pickup, amplifies and squares the pulses, and through a dividing network allows a certain percentage of the pulses to pass to the counter to readout measurements directly in gallons, barrels, etc. The dividing network is established for the turbine meter being used. Figure 5.10 shows a flow totalizer and rate indicator unit. The model shown displays continuous totalization of flow and instantaneous flow rate from a turbine meter. The counter is mounted in a plug-in module that contains the necessary circuitry for totalization and rate indication of one turbine flowmeter.

Turbine Flow Transducers and Probes. Turbine flowmeters have been miniaturized to allow traversing or point measurements as in the case of

Table 5.1
Flow Range Capabilities and Performance Characteristics of the Flanged Model Turbine Meter Shown in Figure 5.7[a]

Flange size (in.)	Flow range GPM	BPH	BPD	Frequency output (Hz)
3/4	1.3-13	1.9-19	45-450	100-1000
3/4	3.2-23	4.6-46	109.7-1097	100-1000
1	6.4-64	9-90	219-2190	100-1000
1 1/2	17.4-174	25-250	600-6000	100-1000
2	33-290	48-480	1141.7-11,417	100-1000
3	60-600	85-850	2057-20,570	50-500
4	107-1071	153-1530	3672-36,720	50-500
6	300-3000	429-4290	10,286-102,860	50-500
8	789-7890	1128-11,280	27,068-270,680	50-500

Repeatability (±% of rate)	Accuracy (±% actual flow)	Temp. range (°F)	Coil inductance (mH)	Coil resistance (ohms)
0.1	0.5	-100 to +225 Std.	400-605	1190-1450
		-100 to +450 Opt.		
		-300 to +800 Opt.		

[a]Courtesy of C-E Invalco, Combustion Engineering Co., Tulsa, Okla.

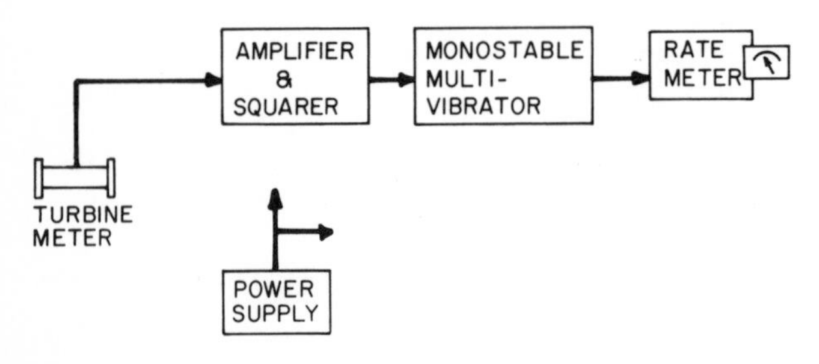

Figure 5.8 Basic setup for the flow rate indicator. (Courtesy of C-E Invalco, Combustion Engineering Co., Tulsa, Okla.)

Table 5.2

Partial Listing of Liquids Measured with Turbine Flowmeters[a]

Anhydrous ammonia	Casillion acid	Fuel oil	Maleic acid	Rosin
Acetic acid	Chlorine propellant	Fish oil	Maleic anhydride	Rust inhibitor
Amines	Coffee extract		Methyl amines	
API condensate	Chocolate slurry	Gasoline	Milk	Silicone tetrachloride
Acetic anhydride	Corn oil	Glacial acetic acid	Milk products	Salt water
Actinium hydroxide	Cellulose slurry	Glycol	Molasses (hot)	Sugar
Arnel	Cyclohexanone		Methanol formaldehyde	Soft drink syrup
Ammonium sulfate	Caustics	Hydraulic oil		Sodium hydroxide
Aromatic naphtha	Chicken fat (hot)	Heat transfer oil	Nitrogen tetraoxide	Sulfuric acid
Aromatic hydrocarbons		Heptane	Nitrogen	Slurries
Apple juice	DMT (dimethyltryptamine)	Hydrogen peroxide	Nitric acid	Synthetic fibers
Adipic acid	Dimethylaniline	Hydrocarbon condensate	Naphtha	
Amyl	Debutanized isopentane	HCN (acid)		Tetraethyl lead
Animal fat (hot)	Detergent		Oil	Transformer oil
Alcohol	Demineralized water	Isocyanate	Oil dispersant	Transformer coolant
Ammonium phosphate	Deionized water	Isopropyl alcohol		Tea extract
Amino phosphonic acid	Distilled water	Isobutanol acetate	p-Dioxane	Toluolene diisocyanate
	Dimethyl ether	Isobutyl	Phthalic anhydride	Toluolene
Benzoic acid	Diesel fuel	Ice cream mix	Phosphoric acid	Trioxane
Brine			Pickle brine	Transmission fluid
Benzene	Ethylacetate	Jet fuel	Phenol	Tetrachloroethylene
Brackish sulfate	Ethylene glycol		Propylene oxide	Toluene
Beer	Ethyl toluene	Kerosene	Propane	Urea
Butanol	Egg yolk		Pentane	
Bisphor A		Lard (hot)	Paint	Vinyl chloride
Butane A		LPG (liquefied petroleum gas)	Polymers	
	Fluorocarbon R-13		Pluoronic L62	Water
Caprolactan	Fluorocarbon R-12			Wax (hot)
Crude oil	Fatty acid	Methanol	Resin	Whiskey
Coconut oil	Fabric dyes	Methylene chloride	Rum	Whey

[a]Courtesy of C-E Invalco, Combustion Engineering Co., Tulsa, Okla.

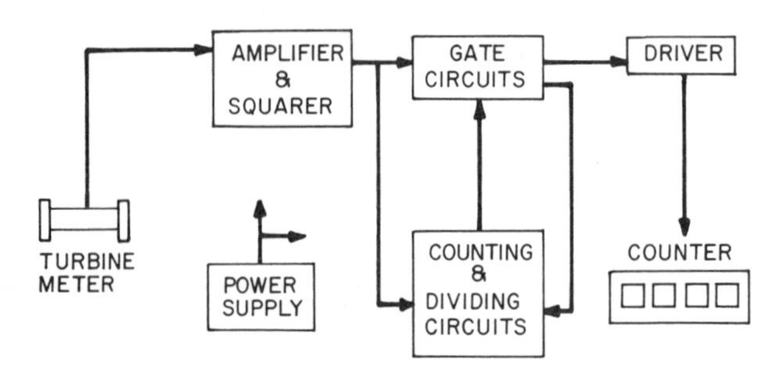

Figure 5.9 Illustrates basic setup for flow totalizer system. (Courtesy of C-E Invalco, Combustion Engineering Co., Tulsa, Okla.)

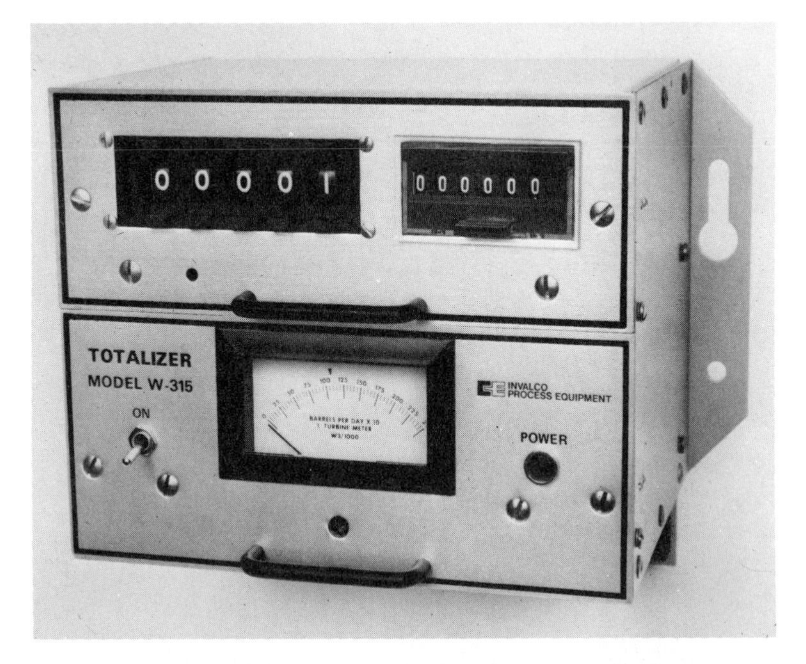

Figure 5.10 Flow totalizer and rate indicator. This unit is equipped with a pulsed analog output that is in proportion to the flow rate, for remote rate indication. Input power requirements are 115 V (ac) or 12 V (dc) at 300 mA. Optional dividing factor switch may be provided on the totalizer module which permits the operator to easily change the calibration of the totalizer from the front panel. (Courtesy of C-E Invalco, Combustion Engineering Co., Tulsa, Okla.)

pitot tubes. One design involves an axial turbine flowmeter element that is mounted on the end of a strut which allows the flowmeter element to be positioned as desired in a large pipe or channel.[57,58]

With this design, the flow velocity is measured by a turbine rotor. The flow velocity profile in a pipe can, for example, be determined by traversing over the pipe diameter with a retractable or hand-held turbine probe. The flow velocity can be directly related to the total flow volume. These systems are specifically designed for large pipe or channel applications and can be coupled with standard readout instrumentation.

Other Designs. There are a variety of flowmeters on the market that are similar to the basic turbine meter. Figures 5.11 and 5.12 show two meters that are specifically designed for metering low liquid flow rates.

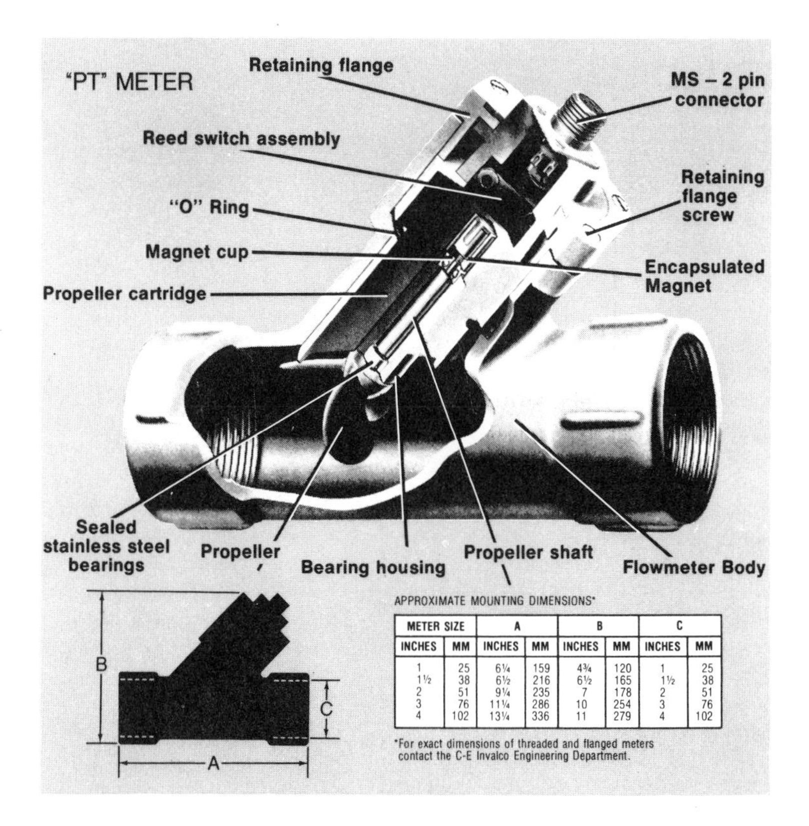

APPROXIMATE MOUNTING DIMENSIONS*

METER SIZE		A		B		C	
INCHES	MM	INCHES	MM	INCHES	MM	INCHES	MM
1	25	6¼	159	4¾	120	1	25
1½	38	6½	216	6½	165	1½	38
2	51	9¼	235	7	178	2	51
3	76	11¼	286	10	254	3	76
4	102	13¼	336	11	279	4	102

*For exact dimensions of threaded and flanged meters contact the C-E Invalco Engineering Department.

Figure 5.11 Illustrates a propeller-type flowmeter. (Courtesy of C-E Invalco, Combustion Engineering, Inc., Tulsa, Okla.)

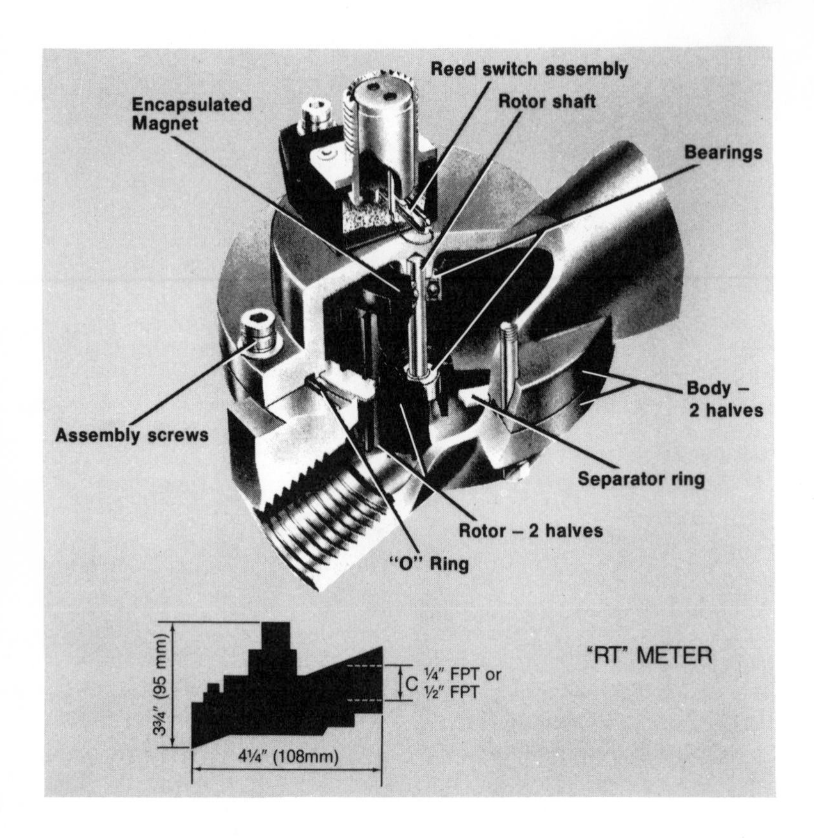

Figure 5.12 Illustrates a rotor-type flowmeter. (Courtesy of C-E Invalco, Combustion Engineering, Inc., Tulsa, Okla.)

Figure 5.11 shows a compact propeller-type flowmeter, designed with a Y-type body to position all components, except a propeller, out of the liquid stream. The main leg of the body forms part of the flow stream, and the other leg (at 45°) contains the propeller cartridge. The propeller is three bladed and designed for maximum clearance between each blade, thus allowing suspended particles to pass. As the propeller goes through each revolution, encapsulated magnets generate pulses through the pickup device (and the number of pulses is directly proportional to the flow rate).[59]

Figure 5.12 illustrates a rotor-type flowmeter that is designed for flow lines with less than 2.5-cm diameter. The manufacturer notes that maximum linearity is achieved at low liquid flow rates. The design includes a ball-bearing-supported rotor for compact design and minimum restrictions to flow. Flow through this meter is unidirectional and mounting may be vertical or horizontal.

6

Magnetic Flow Measuring Systems

6.1 INTRODUCTION

The operation of a magnetic flowmeter is based on Michael Faraday's law of electromagentic induction, which states that the relative motion, at right angles between a conductor and magnetic field, induces a voltage within the conductor. The voltage induced is proportional to the relative velocity of the conductor and the magnetic field.

This chapter discusses some of the fundamental principles which govern the operation of magnetic flowmeters and magentic flow transmitter instrumentation.

6.2 PRINCIPLES OF OPERATION

The operating principle of the magnetic flowmeter is illustrated in Figure 6.1. The flow metering system consists of a primary device, which is the magnetic flowtube, and a secondary device, which is the magnetic flow transmitter. The magnetic flowtube is mounted in the process pipeline, whereas the secondary device can be remote from the flowtube.

The basic components of the flowtube consists of an electrically insulated lined metal or unlined fiberglass-reinforced plastic metering tube with an opposed pair of metal electrodes mounted in the tube wall. A pair of electromagnetic coils are mounted external to the metering tube in a protective casing.[60]

Electric current (ac) is used to excite the coils and generate a magnetic field at right angles to the axis of the process fluid passing through the metering tube.

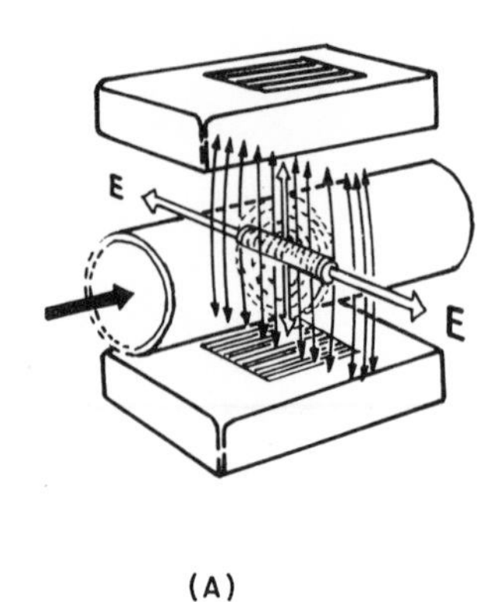

(A)

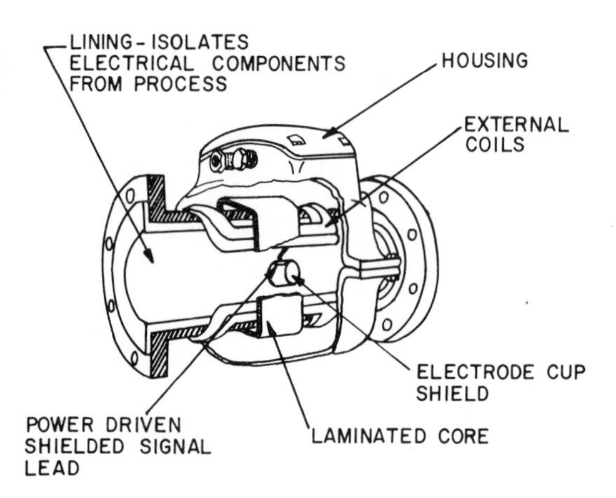

(B)

Figure 6.1 (A) Illustrates the basic operational concept of the magnetic flowmeter. (B) Illustrates the major components of the magnetic flowmeter.

The liquid being metered must be conductive. As the fluid passes through the magnetic field, a voltage (electrical potential) develops across the electrodes. This voltage is directly proportional to the volumetric flow rate.

The voltage signal generated can then be converted in the magnetic flow transmitter to either a standardized dc current signal, which can be transmitted to a suitable receiver unit, or a 0-10 V (dc) output signal.[60]

Magnetic flowmeters were introduced commercially in the 1950s by Foxboro Corp. They have been applied to metering the flow of a wide variety of conductive fluids, ranging from as low as 200 μS (microsiemens)/m (or 2 μmho/cm).[60]

6.3 OPERATING CHARACTERISTICS AND APPLICATIONS

Magnetic flowmeters fall into the classification of obstructionless meters; that is, there is no restriction to the flow of the process fluid. Hence, there is no pressure loss across the meter other than that experienced by an equivalent length of pipe. This is an important advantage to low pressure applications (e.g., fluid systems that are gravity fed and not pumped).

A distinct advantage of this class of flowmeters is their minimum exposure of system components to the actual fluid. As opposed to mechanical meters which have metallic parts exposed to the process fluid, the magnetic flowmeter's only exposed parts are the electrodes and the liner or unlined plastic tube. This minimizes corrosion and/or erosion from problem fluids and liquids containing appreciable suspended solids.

Unlike the venturi meter or differential flow metering devices, the magnetic flowmeter is unaffected by variations in fluid viscosity, temperature, and pressure. It is, however, sensitive to large changes in fluid conductivity and slightly affected by supply frequency and voltage variations. In general, magnetic flow systems feature low power consumption.

Magnetic flowmeters have been applied to a wide variety of process industries. In the chemical process industries, these flowmeters have been widely employed in control loops where reproducibility and flow stability are important. These meters are well suited to operations where flashing is a problem, since no pressure drop occurs across these systems. Magnetic flowmeters are used for metering all types of corrosives, strong acids and bases, and lumpy and viscous materials. Specific metering applications include nitric acid, aqua regia, caustics, polystyrene, butadiene latexes, aluminum chloride, various inorganic salts, ferric acids, and ammonium hydroxide.

In the pulp and paper industry, magnetic flowmeters are used in handling high-temperature black liquors. Examples of applications include measuring stock and pulp flows, whitewater, dyes, additives, bleaching chemicals, lime sludge, clay slurries, liquors, alum, and titanium dioxide slurries.

In the metals mining and refining industries, these meters are used for metering iron ore, copper and aluminum slurries, grinding circuits, and basic mineral slurries. In the steel industry, magnetic flowmeters are employed in a variety of water flow applications in continuous casting and influent/effluent waste treatment systems. The majority of these applications are to abrasive slurries.

In the power industry, these systems are useful for measuring the flow of cooling tower water, recirculating water, carbon slurries, chemical wash, and in SO_2 scrubbing and waste treatment systems.

In the food industry, magnetic flowmeters offer the advantage of clean-in-place design. They are employed in all segments of the food industry, for example, milk standardization, corn processing, aeration control, brewing, and sugar refining.

In the textile industry, magnetic flowmeters are used in chemical metering and waste stream monitoring applications and in measuring the flow of dyes, caustics, resins, and water.

These meters are also widely used in virtually all aspects of wastewater treatment systems, including dredging and mass flow metering.

Magnetic flowmeters are capable of measuring most industrial liquids. Water, acids, bases, slurries, liquids with suspended solids, and industrial wastes are some of the examples already given. The primary limitation of this type of meter is the electrical conductivity of the fluid.

6.4 APPLICATION PRINCIPLES AND METER SPECIFICATION

Meter size, liner material, and materials for the electrodes as well as the specific arrangements for installation will largely depend on the specific fluid to be metered and the process temperature. Specific fluid applications and recommended procedures are discussed.

Liquids with Solids

If the liquid to be metered contains settleable solids that may coat the liner, the meter's transmitter should be sized for normal operating velocities at a minimum of 1.5 m/s. This tends to minimize buildup and allows the coils to be wired in series, thus minimizing the temperature rise.[61]

The transmitter should be installed vertically with the flow upward (this again will minimize coating from solids settling out). The line should be arranged so that when no flow conditions exist the transmitter remains full and thus at a lower temperature than if it were empty.

Liquids that are highly viscous will leave a coating on the tube walls, even after drainage. In such cases, the transmitter should be arranged in the line so that it can be conveniently flushed with a suitable solvent. Also, during lengthy shutdown periods, the tube should be left full of water in order to minimize baking of solids onto the liner surface and/or electrodes.

It should be noted that magnetic flowmeters are far from being trouble-free. Periodic maintenance and cleaning are necessary. For this reason it is often advisable to install the transmitter in a bypass loop or with a "Y" or "T" adjacent to it with removable plug for accessibility.[61]

Abrasive Slurries

The flow transmitter should be installed as in the preceding application with the flow upward. This arrangement tends to equally distribute liner wear.

For liquids that are relatively noncorrosive but are abrasive, an abrasion-resistant material should be selected for the liner. The manufacturer should be consulted for recommendations.

The transmitter should be sized for a maximum fluid velocity of 3 m/s—again, to minimize wear.[61]

The unit should be installed sufficiently downstream of any flow disturbances in order to prevent the fluid stream from impinging on or causing excessive wear to any portion of the liner.

Liquid/Liquid Flows

For liquid/liquid flows, where the fluids are immiscible, the transmitter should be sized for normal operating velocities, no less than 1.5 m/s. The unit should be installed relatively close to a pump or other point where the stream has undergone mixing and the flow is close to being homogeneous. An elbow, for example, will often provide sufficient mixing action to the stream.

Raw Sludge

For applications handling liquids in the form of raw sludge, the unit should be installed in such a manner as to provide easy accessibility for cleaning and repairs. The transmitter should be sized for normal operating velocities in excess of 1.8 m/s.

Temperature Considerations

Excessive or rapid temperature rises in the fluid can adversely effect the meter's accuracy. Permanent damage to the meter and recording instrumentation can occur when excessive temperatures are reached, particularly under no-flow conditions.

The maximum temperature reached in a transmitter under stagnant conditions depends on the initial fluid temperature and the contributed rise due to the energy dissipated by the field coils. The contributed temperature use depends on a number of factors, including the meter size, the field coils connection arrangement (i.e., series or parallel), whether the transmitter is in contact with liquid or not, and the thermal conductivity of the fluid in contact with the tube wall.

Figure 6.2 illustrates the temperature rise with time for a 2-in. magnetic meter. The temperature use above ambient conditions is reported for stagnant conditions, with the transmitter full of water and with air. Figure 6.2 also illustrates the magnitude of the temperature rise experienced for both parallel and series coil connections. As shown, the temperature rise experienced by the parallel-connected coils is substantially greater than with the series-connected arrangements.

The maximum temperature experienced by the transmitter will impact on the maximum permissible temperature for any given liner material as well as the maximum temperature to which the process liquid can be subjected.

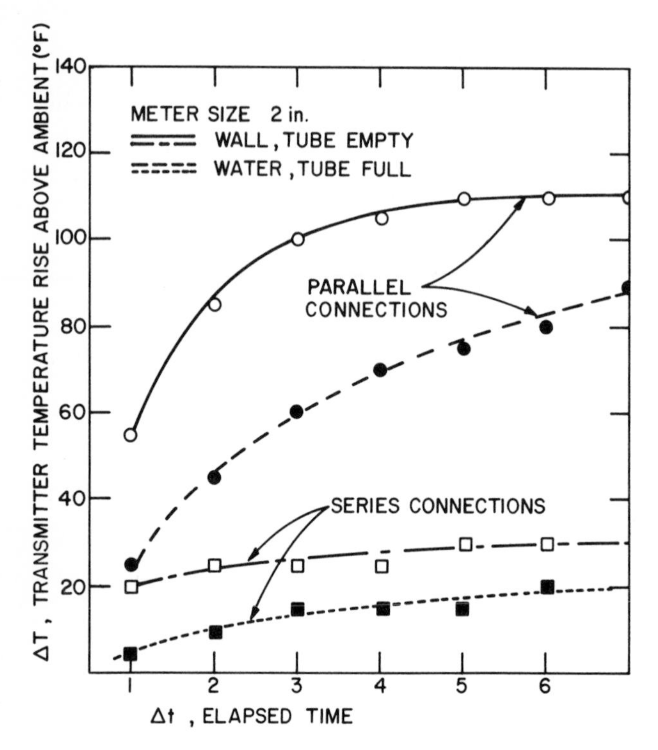

Figure 6.2 Illustrates transmitter temperature rise above ambient conditions. Note that for liquids other than water, the ΔT can be estimated by dividing the temperature rise for water by the product of the specific heat and the specific gravity of the liquid under consideration. The ΔT for a liquid, however, will never exceed that for air. (Data extracted from Ref. 62.)

6.5 MATERIALS SELECTION CONSIDERATIONS

The selection of materials for tube and electrodes will primarily depend on bulk fluid properties, process temperature, and trace constituents in the fluid stream. The user must employ considerable discretion when selecting materials for a specific fluid metering application. Meter manufacturers and/or materials experts should always be consulted for troublesome applications or where there is a problem in selecting the right materials.

The selection of the metal for the electrodes can be a critical decision because of the small size of the electrode and the integrity of sealing that is required. A minimal degree of corrosion is often tolerable for massive structures; however, it can be intolerable on small electrodes. Corrosion rates in excess of 0.0051 cm/year are generally considered unacceptable for magnetic flowmeter electrodes.[62] Temperature considerations must also be weighed in selecting the proper electrode to ensure that it is compatible with the maximum temperature limitation on the liner material.

With regard to liner materials, the manufacturer will supply information on the chemical and abrasion resistance and temperature limitations of various materials. It is important to note that it is not often possible to predict or assess a particular liner material's usefulness in service. Properties are normally reported at one temperature. Properties will, however, vary considerably with temperature (for example, chemical resistance will often decrease with increasing temperature).

The amount and types of contamination that may be present in a fluid stream will also influence liner material selection. The liner material can often be easily selected to meet compatibility requirements with the main constituent in the stream; however, it is usually much more difficult to select a liner that is compatible with a mixture of fluids.

Table 6.1 contains information on the compatibility of various materials for magnetic flowmeter applications. The tabulated data are based on secondary information compiled by one manufacturer. Table 6.1 is useful, however, as a rough guide to materials selection for magnetic flow transmitters.

(Table 6.1 begins on page 114.)

Table 6.1
Rough Guide to Materials Selection for Magnetic Flowmeters[a, b]

Process liquid	Flowtube i.d. material[c]							Electrode metal			
	TFE (T)	Kynar (X)	Kel-F (X)	Polyurethane (A)	Epoxy (Fibercast RB 2530)	Polyester (Atlac 382)	Neoprene (N)	Platinum ±10% iridium (P)	Tantalum (B)	316 SST (S)	Hastelloy C (H)
Acetaldehyde	300	—	250	X	NR	NR	X	P	P	S[1]	P
Acetic acid 10%	300	230	200	X	150	*200	0	P	P	S[1]	P
Acetic acid 25%	300	230	180	X	100	200	0	P	P	S[1]	P
Acetic acid 50%	300	230	180	X	100	160	—	P	P	NR	P
Acetic acid 75%	300	—	120	X	100	160	—	P	P	NR	P
*Glacial 100%	300	150	120	X	100	—	X	P	P	NR	P
Alum	*350	*250	220	—	200	*200	0	P	P	NR	NR
Aluminum chloride	*350	*250	220	—	*200	*200	0	P	P	NR	NR
Aluminum chlorohydrate	*350	*250	—	—	—	200	—	P	P	NR	NR
Aluminum fluoride	*350	*250	—	—	150	75	0	P	NR	NR	NR
Aluminum hydroxide	*350	*250	220	—	150	—	—	P	P	S	NR
Aluminum nitrate	*350	*250	220	—	*200	160	0	P	NR	NR	NR
Aluminum sulfate	*350	*250	220	0	*200	*200	0	P	P	S	S
Ammonium bicarbonate	*350	*250	*250	—	200	160	—	P	P	NR	NR

Ammonium carbonate	*350	*250	*250	—	200	75	0	P	P	NR	NR
Ammonium chloride	*350	*250	*250	—	200	*200	0	P	P	NR	S
Ammonium bifluoride	*350	*250	—	—	—	—	—	P	P	NR	NR
Ammonium fluoride	*350	*250	—	—	150	—	—	P	NR	NR	S
Ammonium hydroxide 10%	*350	*250	*250	0	200	140	0	P	S	S	S
Ammonium hydroxide 20%	*350	*250	*250	0	150	140	0	P	S	S	S
Ammonium hydroxide 30%	*350	*250	*250	0	100	100	0	P	S	S	S
Ammonium nitrate	*350	*250	*250	X	*200	*200	LT	P	P	NR	S
Ammonium persulfate	*350	*250	*250	X	NR	180	0	P	P	NR	NR
Ammonium phosphate	*350	*250	*250	—	150	75	0	P	P	NR	NR
Ammonium sulfate	*350	*250	*250	0	*200	*200	—	P	P	NR	NR
Ammonium sulfide	*350	*250	*250	—	—	—	—	P	P	NR	NR
Antimony pentachloride	*350	*250	—	—	—	70	—	P	P	NR	NR
Antimony trichloride	*350	70	—	—	150	*200	—	P	P	NR	NR
Aqua regia	—	70	120	X	—	NR	X	NR	P	NR	NR
Arsenic acid	*350	*250	*250	X	—	70	0	P	P	NR	NR
Arsenious acid	*350	*250	*250	—	—	180	—	P	P	NR	NR
Barium acetate	*350	*250	*250	—	—	140	—	P	P	NR	NR
Barium carbonate	*350	*250	*250	—	*200	*200	—	P	P	NR	NR
Barium chloride	*350	*250	*250	0	*200	*200	0	P	P	NR	S
Barium hydroxide	*350	*250	*250	0	200	70	0	P	NR	S	NR
Barium sulfate	*350	*250	*250	0	*200	70	0	P	P	NR	NR
Barium sulfide	*350	*250	*250	0	*200	140	0	P	P	NR	NR
Benzene sulfonic acid	—	70	160	—	NR	200	0	P	P	NR	NR
Borax	*350	*250	*250	0	*200	70	0	P	P	NR	NR
Cadmium chloride	*350	*250	*250	—	*200	70	—	S	P	NR	NR

(continued)

Table 6.1
(continued)

Process liquid	Flowtube i.d. material[c]							Electrode metal			
	TFE (T)	Kynar (X)	Kel-F (X)	Polyurethane (A)	Epoxy (Fibercast RB 2530)	Polyester (Atlac 382)	Neoprene (N)	Platinum ±10% iridium (P)	Tantalum (B)	316 SST (S)	Hastelloy C (H)
Calcium bisulfite	*350	*250	175	0	200	180	0	P	P	NR	NR
Calcium carbonate	*350	*250	175	—	*200	*200	—	P	P	S	S
Calcium chlorate	*350	*250	175	—	*200	*200	—	P	P	NR	NR
Calcium chloride	*350	*250	175	0	*200	*200	0	P	P	NR	NR
Calcium hydroxide 15%	*350	*250	175	0	200	160	0	P	P	NR	S
Calcium hydroxide 20%	*350	*250	175	0	200	160	0	P	P	NR	S
Calcium hydroxide 25%	*350	*250	175	0	200	140	0	P	P	NR	S
Calcium hypochlorite	*350	*250	85	—	NR	85	X	P	P	NR	S
Calcium nitrate	*350	*250	175	0	*200	*200	0	P	P	NR	NR
Calcium sulfate	*350	*250	175	0	*200	*200	—	P	P	NR	NR
Chloroacetic acid 25%	300	212	200	—	100	200	—	P	P	NR	S
Chloroacetic acid 50%	300	212	—	—	—	140	—	P	P	NR	S
Chloroacetic acid 100%	300	—	—	—	—	—	—	P	P	NR	S
Chlorosulfonic acid	200	NR	*250	X	—	NR	X	P	P	NR	S

Chromic acid 5%	300	*240	*250	X	150	70	X	P	P	NR	NR
Chromic acid 10%	300	*250	*250	X	150	70	X	P	P	NR	NR
Chromic acid 30%	300	150	*250	X	75	NR	X	P	P	NR	NR
Chromic acid 100%	300	—	—	X	NR	NR	X	P	P	NR	NR
Chromic fluoride	*350	—	—	X	75	—	X				
Chromium sulfate	*350	*250	*250	X	—	140	X	P	P	NR	S
Copper chloride	*350	*250	*250	0	*200	*200	0	NR	P	NR	NR
Copper cyanide	*350	*250	*250	0	—	*200	0	P	P	S	S
Copper fluoride	*350	*250	—	—	*200	—	—	P	NR	NR	NR
Copper nitrate	*350	*250	*250	—	*200	*200	—	P	P	NR	S
Copper oxychloride	*350	*250	—	—	—	—	—	P	NR	NR	NR
Copper sulfate	*350	*250	*250	0	200	*200	0	P	P	S	S
Ferric chloride	*350	*250	*250	0	*200	*200	0	NR	P	NR	S
Ferric nitrate	*350	*250	*250	—	*200	*200	0	P	P	NR	S
Ferric sulfate	*350	*250	*250	—	200	*200	0	P	P	NR	S
Ferrous chloride	*350	*250	*250	—	*200	*200	—	P	P	NR	NR
Ferrous nitrate	*350	*250	*250	—	—	*200	—	P	P	NR	S
Ferrous sulfate	*350	*250	*250	—	200	*200	—	P	P	NR	NR
Fluoboric acid	200	—	—	—	NR	*200	0	P	NR	NR	NR
Fluosilicic acid 10%	200	—	—	—	200	150	0	P	NR	NR	NR
Fluosilicic acid 30%	200	—	—	—	—	70	0	P	NR	NR	NR
Fluosilicic acid 40%	200	—	—	—	—	—	0	P	NR	NR	NR
Fluosulfonic acid	200	—	—	—	—	—	—	—	—	—	—
Formaldehyde	*350	*250	*250	X	150	—	0	P	P	NR	S
Formic acid 10%	*350	*250	180	X	—	150	0	P	P	NR	S
Formic acid 25%	*350	250	180	X	100	—	0	P	P	NR	S

(continued)

Table 6.1
(continued)

Process liquid	Flowtube i.d. material[c]							Electrode metal			
	TFE (T)	Kynar (X)	Kel-F (X)	Polyurethane (A)	Epoxy (Fibercast RB 2530)	Polyester (Atlac 382)	Neoprene (N)	Platinum ±10% iridium (P)	Tantalum (B)	316 SST (S)	Hastelloy C (H)
Formic acid 50%	*350	250	180	X	—	70	0	P	P	NR	S
Formic acid 100%	*350	250	—	X	—	—	0	P	P	NR	S
Glycerin	*350	*250	*250	0	*200	*200	0	P	P	P	P
Hydrobromic acid 50%	*350	*250	200	X	150	160	0	P	P	NR	NR
Hydrochloric (cone)	*350	*250	180	X	200	160	LT	X	P	NR	NR
Hydrocyanic (all)	*350	*250	*250	—	NR	200	LT	P	P	S	S
Hydrofluoric acid 10%	*350	250	120	X	NR	180	LT	P	NR	NR	NR
Hydrofluoric acid 20%	*350	250	120	X	NR	120	LT	P	NR	NR	NR
Hydrofluosilicic 10%	*350	*250	—	—	—	150	LT	P	NR	NR	S
Hydrofluosilicic 35%	*350	*250	—	—	—	70	LT	P	NR	NR	S
Hydrofluosilicic 100%	100	*250	—	—	—	—	LT	P	NR	NR	S
Hydrofluoric 100%	78	212	120	X	NR	—	LT				
Hydrogen peroxide 5%	*350	*250	150	—	150	150	—	P	P	S	S
Hydrogen peroxide 30%	*350	*250	150	—	75	70	—	P	P	S	S

Hydroxy acetic acid 35%	200	—	—	—	—	—	—	P	P	S	S
Hydroxy acetic acid 70%	150	—	—	—	—	—	—	P	P	S	S
Hypochlorous acid 10%	300	—	—	—	200	—	—	NR	P	NR	S
Hypochlorous acid 20%	300	*250	—	—	—	160	—	NR	P	NR	S
Lead acetate	*350	*240	*250	—	*200	*200	LT	NR	P	NR	NR
Lithium chloride	*350	*250	*250	—	—	—	—	P	P	NR	S
Magnesium bisulfite	*350	*250	*250	—	*200	175	—	—	—	S	S
Magnesium carbonate	*350	*250	*250	—	*200	175	—	P	P	S	S
Magnesium chloride	*350	*250	*250	0	*200	*200	0	P	P	NR	S
Magnesium hydroxide	*350	*250	*250	0	*200	—	0	P	NR	NR	NR
Magnesium nitrate	*350	*250	*250	—	*200	—	—	P	P	NR	NR
Magnesium sulfate	*350	*250	*250	—	*200	*200	0	P	P	S	NR
Mercuric chloride	*350	250	150	—	—	*200	0	P	P	NR	NR
Mercurous chloride	*350	250	150	—	—	*200	—	—	—	—	—
Nickel chloride	*350	*250	*250	—	*200	*200	0	P	P	NR	S
Nickel nitrate	*350	*250	*250	—	200	*200	—	P	P	NR	NR
Nickel sulfate	*350	*250	*250	0	—	*200	0	P	P	NR	NR
Nitric acid 10%	*350	*250	*250	X	NR	160	X	P	P	S	S
Nitric acid 15%	*350	*250	*250	X	NR	—	X	P	P	NR	NR
Nitric acid 20%	*350	200	*250	X	NR	—	X	P	P	NR	NR
Nitric acid 40%	*350	200	*250	X	NR	70	X	P	P	NR	NR
Nitric acid 70%	200	120	*250	X	NR	NR	X	P	P	NR	NR
Oxalic acid (saturated)	*350	120	*250	—	*200	*200	LT	P	P	NR	NR
Perchloric acid 70%	300	120	75	—	75	—	0	P	P	NR	NR
Phenol 10%	BP	212	175	X	150	NR	X	P	P	S	S
Phosphoric acid 75%	300	*250	*250	—	*200	*200	—	P	P	NR	NR

(continued)

Table 6.1
(continued)

Process liquid	Flowtube i.d. material[c]							Electrode metal			
	TFE (T)	Kynar (X)	Kel-F (X)	Polyurethane (A)	Epoxy (Fibercast RB 2530)	Polyester (Atlac 382)	Neoprene (N)	Platinum ±10% iridium (P)	Tantalum (B)	316 SST (S)	Hastelloy C (H)
Phosphoric acid 85%	300	230	*250	—	NR	*200	—	P	P	NR	NR
Phos. oxychloride	300	—	—	—	—	—	—	—	—	—	—
Potassium aluminum sulfate	*350	*250	*250	—	—	*200	—	P	P	NR	NR
Potassium bicarbonate	*350	*250	*250	—	*200	70	—	P	P	S	S
Potassium carbonate	*350	*250	*250	—	*200	70	—	P	P	S	S
Potassium chloride	*350	*250	*250	0	*200	*200	0	P	P	NR	S
Potassium dichromate	*350	*250	*250	0	*200	*200	0	P	P	S	NR
Potassium ferricyanide	*350	*250	*250	—	—	*200	0	NR	P	NR	NR
Potassium ferrocyanide	*350	*250	*250	—	—	*200	—	NR	P	NR	NR
Potassium hydroxide 10%	200	200	250	LT	200	150	0	P	NR	NR	NR
Potassium hydroxide 20%	200	200	250	LT	—	70	0	P	NR	NR	NR
Potassium hydroxide 45%	200	212	250	LT	—	—	0	P	NR	NR	NR
Potassium nitrate	*350	*250	*250	0	*200	*200	0	P	P	NR	NR
Potassium permanganate	200	*250	250	—	150	*200	—	P	P	NR	NR

Potassium persulfate	*350	*250	250	—	—	*200	—	P	P	S	S
Potassium sulfate	*350	*250	250	0	150	*200	0	P	P	S	NR
Silver nitrate	*350	*250	*250	0	*200	*200	0	P	P	NR	S
Sodium acetate	*350	*250	*250	X	200	*200	LT	P	P	NR	NR
Sodium bicarbonate	*350	*250	*250	—	*200	140	0	P	P	S	S
Sodium bisulfate	*350	*250	*250	—	*200	*200	—	P	P	NR	NR
Sodium bisulfide	*350	*250	*250	—	—	—	—	P	P	S	NR
Sodium bisulfite	*350	*250	*250	—	—	*200	0	P	P	NR	S
Sodium borate	*350	*250	*250	—	—	—	0	P	P	NR	NR
Sodium bromide	*350	*250	*250	—	200	*200	—	P	P	NR	NR
Sodium carbonate	*350	*250	*250	—	*200	160	—	P	P	S	S
Sodium chlorate	*350	*250	*250	—	—	*200	—	P	P	NR	NR
Sodium chloride	*350	*250	*250	0	*200	*200	0	P	P	NR	NR
Sodium chlorite	*350	*250	*250	—	—	150	—	S	P	NR	NR
Sodium chromate	*350	*250	*250	—	—	*200	—	P	P	NR	NR
Sodium cyanide	*350	*250	*250	—	*200	*200	0	NR	P	S	—
Sodium dichromate	*350	*250	*250	—	*200	*200	—	—	—	—	—
Sodium ferricyanide	*350	*250	*250	—	—	*200	—	NR	S	NR	—
Sodium ferrocyanide	*350	*250	*250	—	*200	—	—	NR	S	NR	—
Sodium fluoride	*350	*250	—	—	*200	—	—	P	NR	NR	NR
Sodium hexametaphosphate	*350	*250	*250	—	—	—	—	—	—	—	—
Sodium hydrosulfide	*350	*250	*250	—	—	—	—	P	P	S	S
Sodium hydroxide 5%	*350	212	*250	LT	200	200	0	P	NR	S	NR
Sodium hydroxide 10%	*350	212	*250	LT	200	160	0	P	NR	S	NR
Sodium hydroxide 25%	*350	212	*250	LT	200	140	0	P	NR	S	NR
Sodium hydroxide 50%	*350	212	*250	LT	200	70	0	P	NR	S	NR

(continued)

Table 6.1
(continued)

Process liquid	Flowtube i.d. material[c]							Electrode metal			
	TFE (T)	Kynar (X)	Kel-F (X)	Polyurethane (A)	Epoxy (Fibercast RB 2530)	Polyester (Atlac 382)	Neoprene (N)	Platinum ±10% iridium (P)	Tantalum (B)	316 SST (S)	Hastelloy C (H)
Sodium hypochlorite 5%	BP	*250	75	X	NR	—	LT	P	P	NR	S
Sodium hypochlorite 10%	—	*250	—	X	NR	—	LT	P	P	NR	S
Sodium hypochlorite 15%	—	*250	—	X	NR	—	LT	P	P	NR	S
Sodium hypochlorite 20%	200	*250	—	X	NR	—	LT	P	P	NR	S
Sodium nitrate	*350	*250	*250	—	*200	*200	0	P	P	S	NR
Sodium nitrite	*350	*250	*250	—	—	*200	—	P	P	NR	NR
Sodium silicate	*350	*250	*250	—	150	*200	0	P	P	NR	NR
Sodium sulfate	*350	*250	*250	0	*200	*200	0	P	P	NR	NR
Sodium sulfide	*350	*250	*250	—	—	*200	0	P	P	NR	NR
Sodium sulfite	*350	*250	*250	—	200	*200	0	P	P	S	NR
Sodium tetraborate	*350	*250	*250	—	—	—	—	P	P	S	S
Sodium thiosulfate	*350	*250	*250	0	150	—	0	NR	NR	NR	P

Stannic chloride	*350	*250	*250	—	200	*200	0	P	P	NR	NR
Stannous chloride	*350	*250	*250	—	—	*200	0	P	P	NR	NR
Sulfamic acid	300	*250	—	—	—	*200	—	P	P	—	—
Sulfuric acid 10%	*350	*250	*250	LT	*200	*200	LT	P	P	NR	NR
Sulfuric acid 25%	*350	*250	*250	0	150	*200	0	P	P	NR	NR
Sulfuric acid 50%	*350	*250	*250	0	100	*200	0	P	P	NR	NR
Sulfuric acid 80%	*350	212	*250	X	NR	70	X	P	P	NR	NR
Sulfuric acid 100%	300	150	*250	X	NR	—	X	P	NR	NR	NR
Sulfurous acid 10%	*350	212	—	0	200	—	LT	P	P	NR	NR
Trisodium phosphate	*350	*250	250	—	150	70	—	P	P	S	S
Zinc chloride	*350	*250	*250	—	*200	*200	0	P	P	NR	NR

[a]Courtesy of Foxboro Corp., Foxboro, Mass.

[b]Key: X, NR, not recommended; P, preferred material (virtually unlimited life); S, satisfactory material (reasonable service life); LT, low temperature use only (~100° F); 0, compatible to rated tube temperature.

[c]A maximum recommended temperature marked with an asterisk (*) indicates that the transmitter construction is the limiting factor as opposed to the tube i.d. material in the particular liquid being the limiting factor. Maximum recommended temperatures of process liquids in transmitters, unless otherwise noted, are as follows: Kynar or Kel-F, 250° F; urethane, 160° F; neoprene, 150° F; tetrafluoroethylene (TFE), 350° F; and epoxy- or polyester-reinforced fiberglass (unlined), 200° F.

7

Ultrasonic Flow Measurement Systems

7.1 INTRODUCTION

Doppler transducer or ultrasonic flow measuring systems meter fluid flow by employing the Doppler frequency shift of ultrasonic signals reflected from discontinuities in the flowing liquid. This class of flowmeters offers the following advantages:

1. Noninvasive flow measurement
2. No moving parts
3. No pressure loss

This chapter discusses the basic principles of operation of the Doppler flowmeter, specifically concerning applications to flow measurement of liquids and slurries in pipes. Applications of other ultrasonic techniques to open channel flow are briefly discussed at the end of the chapter.

7.2 MEASUREMENT PRINCIPLE (DOPPLER)

Flow measurement is achieved by using the Doppler frequency shift of ultrasonic signals that are reflected from various forms of discontinuities in the liquid stream. These discontinuities can be in the form of suspended solids, bubbles, or interfaces generated by turbulent eddies in the flow stream.

Figure 7.1 illustrates the principle of operation. The meter or "sensor" is mounted on the outside of the pipe, and an ultrasonic beam from a piezoelectric crystal is transmitted through the pipe wall into the fluid at an angle to the flow stream.

Signals reflected off of flow disturbances are detected by a second piezoelectric crystal located in the same sensor. Transmitted and reflected signals are compared in an electric circuit. The frequency shift is

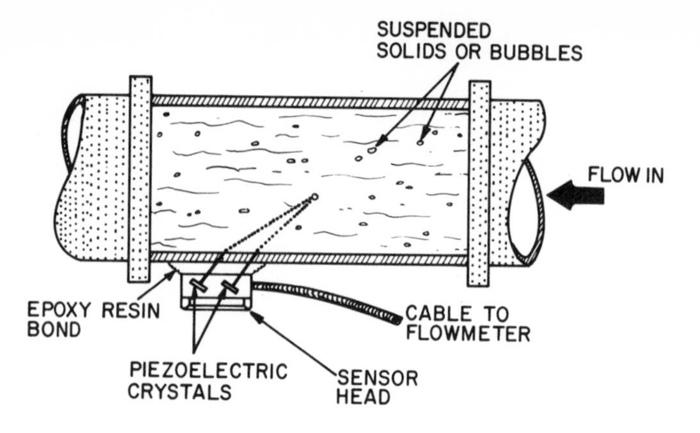

Figure 7.1 Application of the Doppler flowmeter to metering liquids in pipe flow. (Courtesy of Bestobell Mobrey Limited, Slough, Bucks, England.)

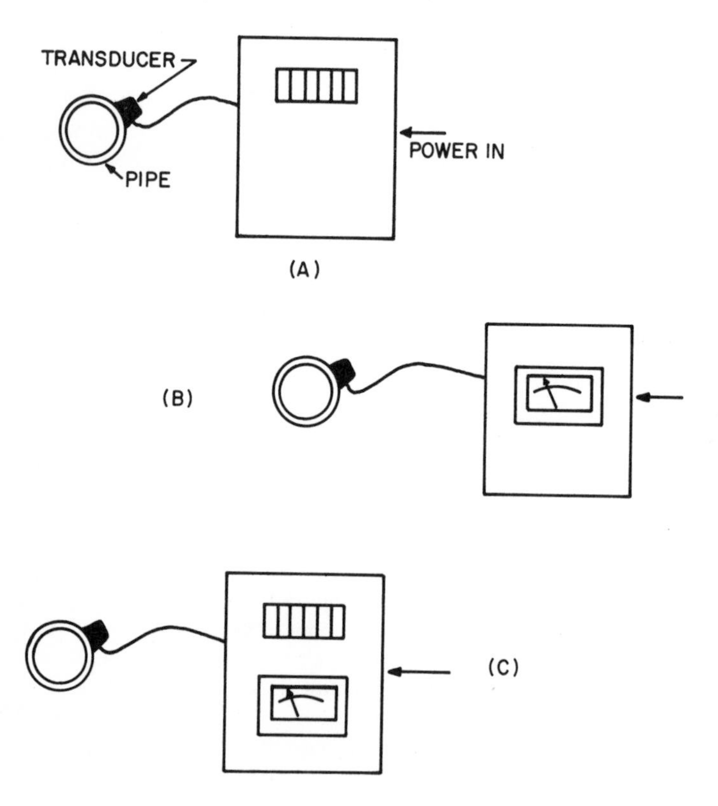

Figure 7.2 The Doppler transducer flowmeter can be coupled with: (A) a controller that consists of a totalizer which indicates the volume metered; (B) a flow indicator that records the rate of mass flow or velocity; or (C) a flowmeter totalizer that indicates the rate of flow or velocity.

proportional to the flow velocity. The electronic circuit is employed for the addition of factors to take into account pipe and liquid constants prior to generating an analog signal to feed a meter and totalizer or batch counter (refer to Figure 7.2).[63] Figure 7.3 shows a Doppler transducer and power pack. Also refer to Figures 7.4 and 7.5, which show a miniaturized Doppler sensor and flow indicator, and a flow velocity meter, respectively. The electronic units shown in Figures 7.3 through 7.5 are factory calibrated to provide readouts of flow rate (i.e., meters per second). Internal adjustments can be made to accommodate variations generated from different flow profiles or site conditions. Usually other internal adjustments enable flow speed readings to be translated into suitable volume units for meter or totalizer readouts, calculated in accordance with pipe sizes.

Factory calibration is generally good to within 5%, based on information supplied on pipe material and size, site conditions, fluid composition, and temperature. On-site flow calibration can, however, provide measurement accuracy to within 1% of the actual flow.[63] The amplitude of reflected signals will largely depend on the number of particles or discontinuities present in the flow stream.

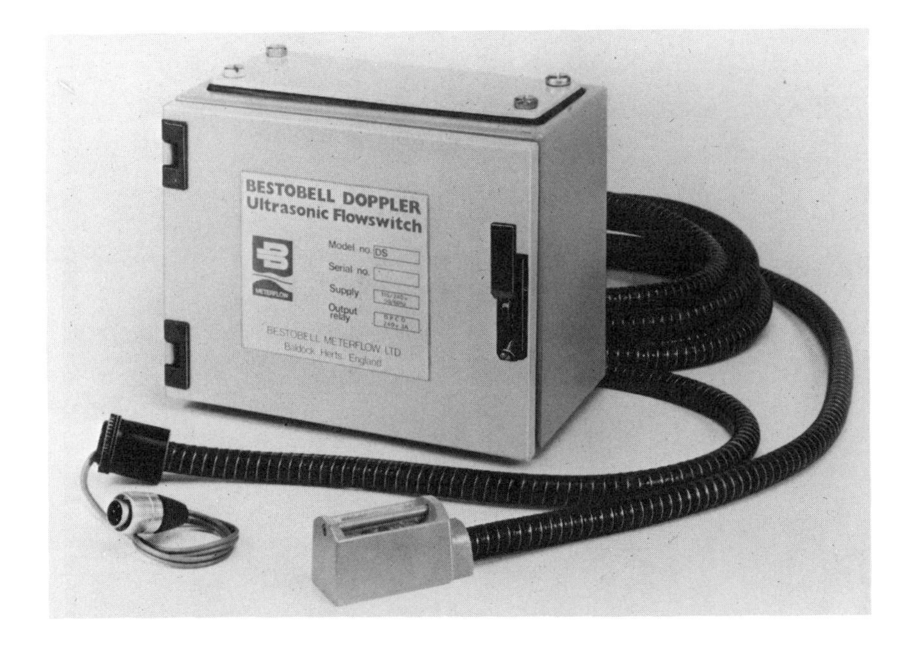

Figure 7.3 Doppler transducer flow measurement system. (Courtesy of Bestobell Mobrey Limited, Slough, Bucks, England.)

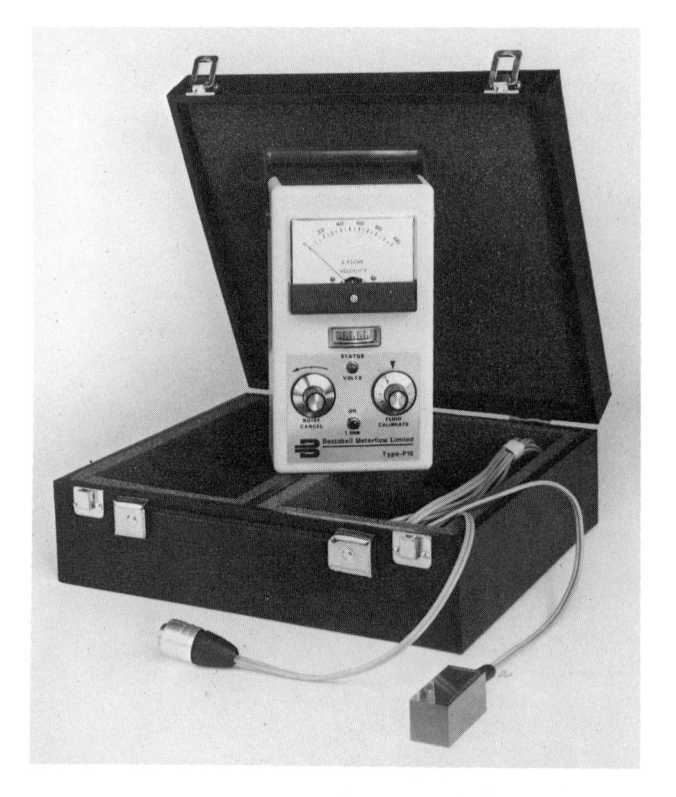

Figure 7.4 Miniaturized Doppler transducer and flow velocity indicator. (Courtesy of Bestobell Mobrey Limited, Slough, Bucks, England.)

Figure 7.5 A flow velocity indicator. (Courtesy of Bestobell Mobrey Limited, Slough, Bucks, England.)

7.3 ADVANTAGES AND OPERATING CHARACTERISTICS

Doppler flowmeters have several advantages. They allow noncontact flow measurements by means of an externally mounted transducer. The unit is relatively easy to install and align, permitting flow measurement with no alterations to piping, no downtime, and no pressure loss.

There is no upper size limit on pipe diameters; hence, capital costs are the same regardless of the application. In addition, this flow measuring system is adaptable to a fairly wide range of piping materials. The system works well with metal, plastic, and other homogeneous piping materials.

The Doppler system is applicable to a variety of problem liquids. Specific examples include slurries, acids, viscous fluids, and various liquids containing suspended solids.

The Doppler system does have some limitations, however. The principal limitations are as follows:

1. Liquids to be metered must have an excess of approximately 2% suspended solids by volume.
2. Liquid linear velocities must exceed 0.15 m/s.
3. The transducer head must be maintained at a temperature below 83° C.
4. Piping material must be of a homogeneous composition (i.e., materials composed of more than one substance, such as concrete, reinforced concrete, vitreous clay, or asbestos cement cannot be used with this method).
5. Pipe wall thickness cannot be greater than 1.91 cm.

The Doppler system is only applicable to full pipe flow. Multiphase and some types of multicomponent flows cannot be metered by this method. Probably the most serious limitation on this method is that it can only meter liquid flows. The Doppler method will not work on solids or gases.

Doppler flowmeters are generally low-maintenance systems. Since the sensor is mounted on the outside of an existing standard plant pipework (i.e., there is no intrusion to the flow), the meter cannot become fouled by dirt, corrosion, or suspended solids. In fact, the presence of dirt or suspended solids provides excellent reflectors for the ultrasonic signal to give a high-amplitude Doppler signal.

7.4 TRANSDUCER MOUNTING AND INSTALLATION CONSIDERATIONS

The ideal location for the transducer, as recommended by manufacturers, is on a vertical pipe.[63,64] If the sensor has to be mounted on a horizontal pipeline, a 45° or 135° position from the pipe bottom is recommended (refer to Figure 7.6).

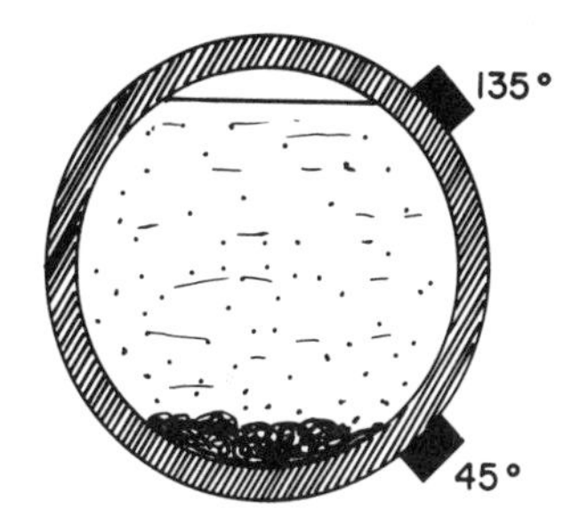

Figure 7.6 Proper location of Doppler transducer for a horizontal flow application.

The transducer should be located roughly ten diameters of straight pipe downstream of any flow disturbances (i.e., pumps, tees, elbows, reducers, etc.). If the sensor is located near such disturbances, calibration adjustments are often necessary which usually affect the linearity of the instrument over a wide flow span.

The sensor is normally bonded to the outside pipe wall with a slow hard-setting epoxy resin. The surface to which the transducer is to be bonded should be thoroughly cleaned of any oil, dirt, paint, rust, etc. Epoxy is normally applied to the transducer face and the unit firmly pressed onto the pipe surface. Care must be taken to avoid cracks or air holes in the epoxy. The sensor should be evenly attached to the pipe surface as a wedge-shaped bond—or excessive epoxy will cause the signal to be dispersed, causing erroneous measurements. During the bonding process the sensor is positioned parallel to the pipe axis and maintained stationary by a mechanical clamp.

Note that it is often advisable to select the permanent installation site after site testing is performed. Initial site testing can be done by employing silicon grease as a couplant between the pipe wall and sensor.

7.5 APPLICATION OF ULTRASONIC TECHNIQUES TO OPEN CHANNEL FLOW

Ultrasonic flow measuring techniques have been successfully applied to open channel flow. The principle of operation is different from that of the Doppler flowmeter. Ultrasonic open channel flow measuring systems translate ultrasonic water depth measurements to a linear readout of flow rate. A computer integrator is usually employed to translate depth measurements to an indication of flow rate, to totalize flow, and to control flow-proportional composite sampling.[65] Figure 7.7 illustrates one type of

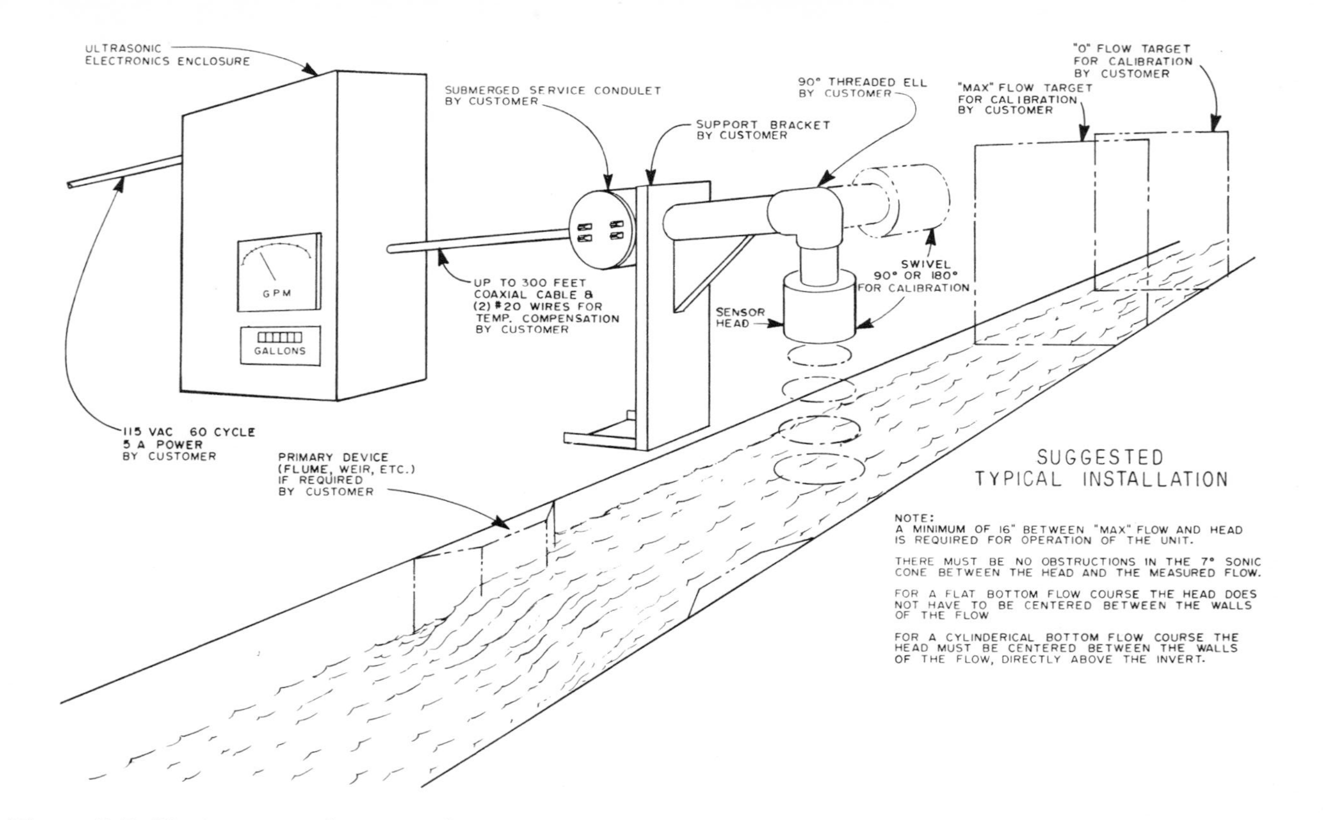

Figure 7.7 Illustrates an ultrasonic flow measuring system as applied to open channel flow. Such an arrangement can be applied to industrial and municipal flow applications. (Courtesy of Cochrane Environmental Systems, Pa.)

installation to metering the flow in open channels. The unit shown can be programmed to match the characteristics of all standard flumes, weirs, and nonrestricted free-flowing pipes and channels where flow rate is always the same at a given depth.

The front face of the sensor (see Figure 7.7) must be positioned so that the signal beam is a 7° cone projected from the edges of the front face, encompassing the area of the flow being monitored. The sensor can be mounted up to 152 m from the electronic system (this varies with the model). Special cable connections between the sensor and recording instrument are necessary for temperature compensation.[66]

System calibration is accomplished by positioning targets to represent maximum and zero flow with reference to the ultrasonic sensor face. Span and zero potentiometers are adjusted until calibration has been achieved.

8

Mass Flow Measurement

8.1 INTRODUCTION

In industry, the need for highly accurate flow measurement becomes very important when processes must be optimized and waste streams minimized. The need is especially acute when dealing with costly chemicals or valuable petrochemicals and petroleum products. Mass flow measurement techniques allow this close control. In addition, because energy utilization is critical, mass flow measurement techniques have been found to be one of the most effective methods of optimizing fuel consumption.

There are a number of advantages to using mass flow measurement. One is the significant reduction in the number of measurements that may be required and their associated error. Process parameters such as pressure, temperature, and specific gravity do not have to be considered in the measurements; hence, data-handling requirements are often simplified.

This chapter discusses the theory of operation of mass flowmeters. Operating characteristics and limitations are included.

The latter part of the chapter discusses the basic principles of anemometry techniques.

8.2 THEORY OF OPERATION

In principle, linear mass flowmeters can be considered hydraulic equivalents of the electrical Wheatstone bridge. The meter is schematically shown in Figure 8.1. As shown, the meter consists of four matched orifices arranged in a hydraulic bridge. The pump shown across the bridge serves to create a constant recirculating reference flow or hydraulic potential within the bridge. Since the orifices are matched, the recirculating flow q becomes equally distributed in the arms of the bridge.

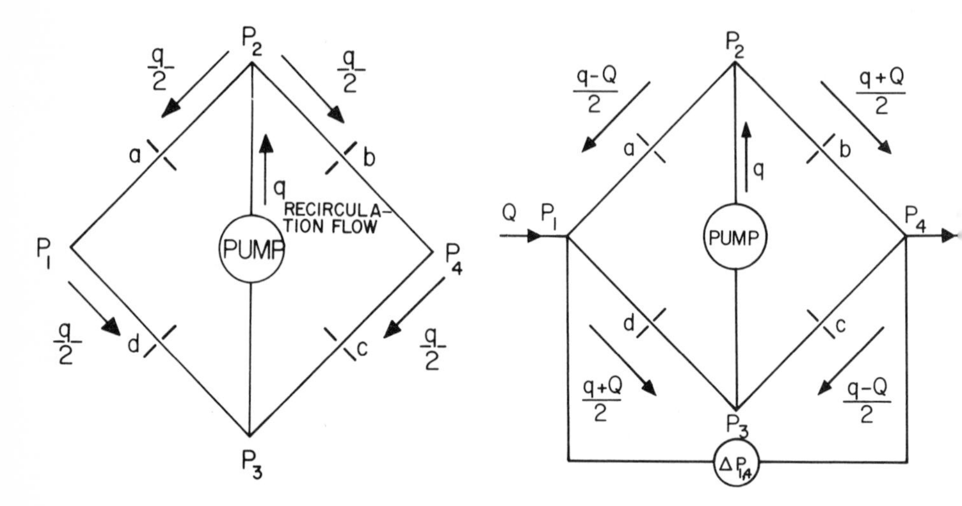

Figure 8.1 Schematically shows the operation of the linear mass flowmeter. (Courtesy of Flo-Tron, Inc., Paterson, N.J.)

Low Flow Measurement

Linear mass flowmeters have been widely used in metering relatively low flow rates. In Figure 8.2 a fluid stream is imposed across the bridge. If $q > Q$, then the flow through orifices b and d will be

$$q_{bd} = \frac{q + Q}{2} \tag{8.1}$$

and the flow through orifices a and c will be

$$q_{ac} = \frac{q - Q}{2} \tag{8.2}$$

Recall from Chapter 3, that the classic equation for flow through a single orifice is

$$Q = CA\sqrt{\frac{\Delta P}{\rho}} \tag{8.3}$$

where C is the orifice discharge coefficient; A is the orifice area; ΔP is the pressure drop across the orifice; and ρ is the fluid density.

Hence the flow versus pressure drop relationship for orifice b can be written as:

$$\frac{q + Q}{2} = C_b A_b \sqrt{\frac{P_2 - P_4}{\rho}} \tag{8.4}$$

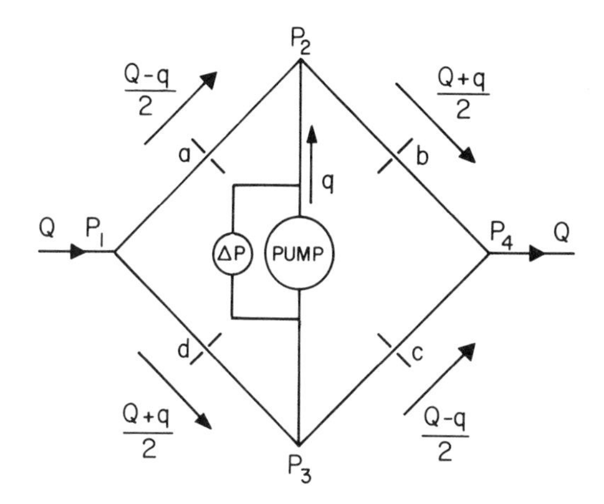

Figure 8.2 Hydraulic bridge with output signal for Q > q. (Courtesy of Flo-Tron, Inc., Paterson, N.J.)

and for orifice a

$$\frac{q - Q}{2} = C_a A_a \sqrt{\frac{P_2 - P_1}{\rho}} \tag{8.5}$$

By squaring Eqs. (8.4) and (8.5), the following is obtained:

$$\frac{q^2 - 2Qq + Q^2}{4} = C_b^2 A_b^2 \frac{P_2 - P_4}{\rho} \tag{8.6}$$

$$\frac{q^2 - 2Qq + Q^2}{4} = C_a^2 A_a^2 \frac{P_2 - P_1}{\rho} \tag{8.7}$$

Since it has been assumed earlier that the orifices are balanced or matched (and for practical purposes),

$$C_b A = C_a A_a = CA \tag{8.8}$$

Then, combining Eqs. (8.6) and (8.7), the following expression is derived:

$$Qq = C^2 A^2 \frac{P_1 - P_4}{\rho} \tag{8.9}$$

Equation (8.9) can be rearranged in terms of the pressure differential across the hydraulic bridge

$$\Delta P_{1,4} = K\dot{m} \tag{8.10}$$

where $K = q/(CA)^2$ = a constant, and m is the process stream mass flow rate $(Q\rho)$.

High Mass Flow Application

The principle of the hydraulic Wheatstone bridge has also been applied to handling fairly large ranges of flow conditions. As in the previous case, the output signal is a differential pressure signal, linear and proportional to the true mass flow.

For the case where $q < Q$, the flow through orifices b and d and through a and c will be given by Eqs. (8.1) and (8.2), respectively [refer to Figure 8.2, which is the normal arrangement for high flow application].

The flow-pressure drop expression for orifice d is given as

$$\frac{Q + q}{2} = C_d A_d \sqrt{\frac{P_1 - P_3}{\rho}} \tag{8.11}$$

and for orifice a

$$\frac{Q - q}{2} = C_a A_a \sqrt{\frac{P_1 - P_2}{\rho}} \tag{8.12}$$

Following the derivation for low mass flow,

$$Qq = C^2 A^2 \frac{P_2 - P_3}{\rho} \tag{8.13}$$

$$\Delta P_{2,3} = KQ\rho \tag{8.14}$$

where K is defined as before.

Equation (8.14) allows the mass flow rate to be computed as a function of the pressure rise across the recirculating pump for the case where $Q > q$.

The selection of whether the recirculation rate q is greater or less than the process stream flow depends on the magnitude of Q, desired rangeability, and other factors.[67,68]

8.3 THE BASIC SYSTEM AND OPERATING PROBLEMS

One manufacturer's low mass flowmeter unit is shown in Figure 8.3. The unit shown is capable of measuring extremely low flow rates down to 0.1 lb/h with extreme accuracy [to within ±0.5% of the reading (±0.03% of full scale)]. This particular model was designed to assist in work on automobile engine emission control and fuel economy.

Figure 8.3 A liquid mass flowmeter unit for measuring extremely low flow rates. This unit has been specifically designed for application to engine emission control and fuel economy. (Courtesy of Flo-Tron, Inc., Paterson, N.J.)

Integrated with the flowmeter is a small heat exchanger. Fuel normally comes from a temperature-controlled supply. Measured quantities of fuel pass through the flowmeter while an additional amount of fuel is passed through the heat exchanger and serves as a coolant. The recirculating pump is located on the other side of the heat exchanger. This arrangement allows the orifice housing to be maintained at a constant temperature, independent of the surroundings.[68]

The specifications for the model shown in Figure 8.3 are given in Table 8.1 to illustrate the range and versatility of these systems.

The four orifices are contained in an orifice housing which has the differential pressure ports and influent and effluent ports. Standard differential pressure transducers can be employed to sense the meter signal. (The recirculating pump in the unit shown in Figure 8.3 is a gear pump being driven at constant speed by a synchronous motor. This provides the necessary constant volume reference flow through the bridge arms.)

Units having low flow capability are generally plagued with a series of problems. Therefore, care should be taken in selecting a particular system. Major problems include signal noise, signal drift due to temperature changes, and sensitivity.

In general, signals from all systems have some noise associated with them. Noise reduction is usually achieved by filtering the signal or by down response time, which is not desirable when metering low flows. Another

Table 8.1
Specifications of a Low Flow Rate Meter (Model 10M Microflow Meter)[a]

Flow range (lb/h)	0-10
Accuracy	±0.5% (±0.03% of full scale)
Repeatability	±0.25% (±0.03% of full scale)
Accuracy and repeatability span	50/1 with 0.1 lb/h minimum flow
Liquids	Gasoline, diesel, kerosene, similar hydrocarbons
Meter pressure drop (in H_2O/lb/h)	1.2
Operating pressure, maximum (psig)	200
Hazardous location	Explosion proof, Class 1, Group D
Heat exchanger fuel coolant flow (lb/h)	150 (minimum)
Materials of construction	Aluminum, steel, bronze, stainless, Buna N
Size (in.)	28(L) × 10-3/16(W) × 11-7/16(H)
Weight (lb)	68
Internal fluid volume (points)	2
Response	1 Hz or 1/16-s step response

[a]Courtesy of Flo-Tron, Inc., Paterson, N.J.

approach to noise reduction is through the use of the low flow recirculating pump on these systems. This reduces turbulent flow noise in the meter, a primary source. Systems are usually equipped with special screened baffles within the orifice housing to further dampen turbulent noise.

Drift resulting from transient fluid temperature variations is another critical parameter affecting low flow measurement. Temperature changes can be attributed to variations in the process stream or ambient conditions. Changes in temperature will consequently impact on fluid density, causing transient flow conditions to arise. That is, temperature changes can cause a flow transient, which is directly proportional to the magnitude of the time rate of temperature change and the volume of fluid affected.[68, 69] Although for large flows this effect is often neglected, the impact on microflow measurements can be great. Because of this, the fluid temperature must be stabilized so that the flow transient is minimized. This is done by the use of the aforementioned heat exchanger.

For most applications the coolant can be water or in some cases fuel. The important criterion for the specific coolant used is its own temperature stability.

8.4 CALIBRATION

The weigh scale and timer method can be used to calibrate this type of flowmeter. The time of flow of a known amount of fluid can be measured by employing a balanced scale to start and stop an electronic timer. The weight of fluid collected divided by the recorded time interval gives the mass flow rate. High precision scales can be employed for extreme accuracy.

For volatile liquids (for example, gasoline), the amount of evaporation from the collection vessel can cause significant errors in the calibration data. Evaporation may be controlled by enclosing the collection vessel so that the volume above the liquid surface does not circulate to the surroundings. The enclosed volume will eventually reach a saturated condition and usually variations in the vapor content during a run can be considered negligible. Note that the flow tube feeding the collection tank or beaker must discharge within the enclosed vessel to minimize vaporization losses.

The weigh scale and timer can be mounted on a calibration stand along with a reservoir. The reservoir contains the liquid for calibrating the meter as well as the coolant for the heat exchanger.

Typical output data are shown in Figure 8.4. Recall that the hydraulic bridge produces an output of differential pressure which is both linear and proportional to the true mass liquid flow rate. The primary advantage with this type of flowmeter is that measurements are essentially unaffected by fluid temperature variations and (consequently) density. In addition, these meters are unaffected by changes in viscosity. The meter can be set to measure flow rate in direct units of mass flow (e.g., lb/h, kg/s).

Figure 8.5 shows a typical calibration curve. The error band shown is ±0.5% of the reading (with ±0.03% of full scale).

It should be noted that the systems described are true mass flowmeters. That is, this class of meters gives the mass rate of flow regardless of the fluid (see data in Figure 8.4). In the automotive industry, mass flow is important because fuel energy is related to the mass of the fuel (i.e., Btu/lb). As such, if volume flow were the measuring parameter, specific gravity would also have to be measured in order to convert volume flow to mass flow. Specific gravity measurements introduce additional error into flow calculations. Furthermore, fluid specific gravity may vary from batch to batch for the same fluid as well as with temperature. As an example, Masnik[68] notes that a 2° F change in fuel temperature can cause a 0.1%

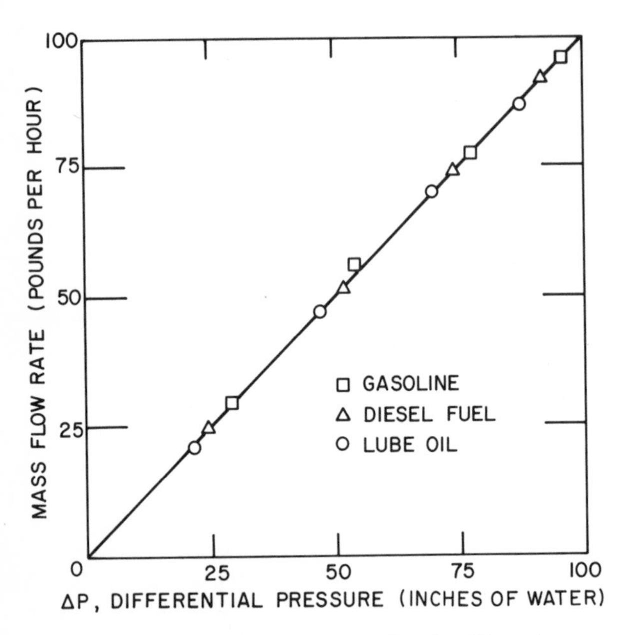

Figure 8.4 Typical output signal and calibration curve from mass flowmeters. (Courtesy of Flo-Tron, Inc., Paterson, N.J.)

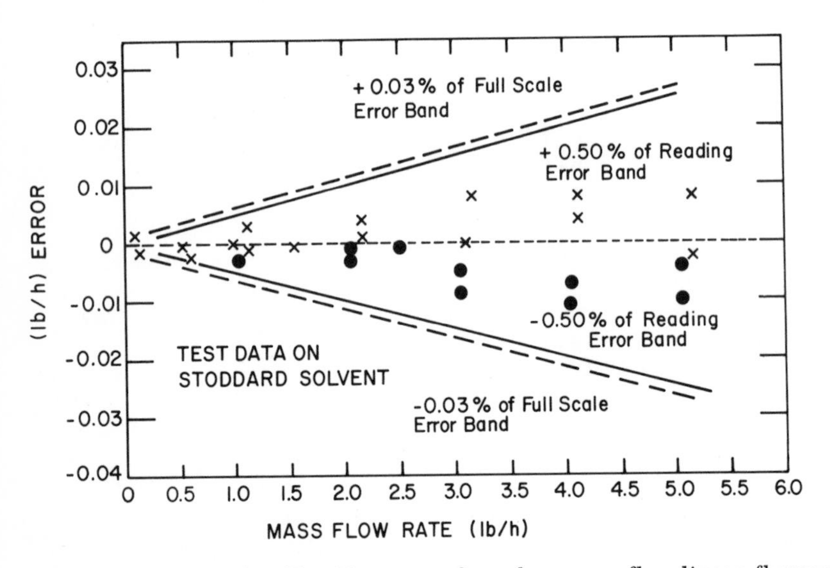

Figure 8.5 Typical calibration curve for a low mass flow linear flowmeter. The plot shows the error band associated with the recorded data. (Courtesy of Flo-Tron, Inc., Paterson, N.J.)

specific gravity variation in gasoline. Under constant volumetric flow conditions with this temperature variation, the mass flow would experience a 0.1% change. In some applications where flow rates as low as 0.1 lb/h are experienced, this error can be significant. Direct measurement of mass flow is preferable to measurement of volumetric flow in order to minimize all possible sources of measuring error.

8.5 MASS FLOW TRANSDUCERS

Mass flow transducers have been applied to mass flow measurements of gases and liquids. These systems employ a heat transfer principle to provide direct reading of a true mass flow rate.

A heated-sensor technique is used as the basic principle. A hot sensor anemometer is used to instantaneously measure fluid-flow parameters by sensing the heat transfer between an electrically heated sensor and the flow medium.

The anemometer is essentially a Wheatstone bridge with one arm of the bridge comprising the probe or heated sensor (refer to Figure 8.6). When the system is operated as a constant temperature anemometer, the bridge circuit supplies heating power to the sensor to raise its temperature above ambient conditions. When the sensor is exposed to a flow stream (which can be either a liquid or gas) the flow conducts heat away from the sensor. The rate at which heat is lost from the sensor is a direct measure of the fluid velocity. The rate of heat removed is indirectly measured by the instantaneous power surge from the circuit that is required to maintain the probe at one temperature. Mathematically, the heat transfer from the heated element (in this case let us say a wire) as a function of the fluid velocity can be expressed as

$$I^2R = (R - R_e)[\alpha + \beta(\rho u)^{1/n}] \tag{8.15}$$

Here, α and β are constants that depend on the sensor dimensions and on fluid properties such as thermal conductivity, viscosity, and specific heat. The other parameters in Eq. (8.15) are as follows:

ρ = density of fluid

u = fluid velocity

n = empirical constant that is dependent on the velocity range and fluid

R = resistance of the hot element

R_e = resistance of the element under ambient conditions

I = electrical current in the heated sensor

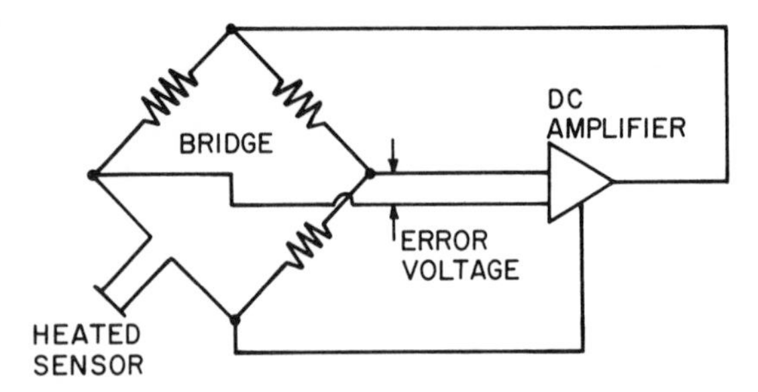

Figure 8.6 Illustrates a constant-temperature anemometer circuit. Systems are also designed to operate in constant current mode.

Equation (8.15) is nonlinear. The output signal can be in terms of either I or power (I^2R). The basic variable that is measured is the mass flow (ρu) per unit area. From this measurement mass flow can be obtained by ρuA, where A is the area of the flow path.

Figure 8.7 shows a typical output curve for a heated sensor exposed to a moving fluid stream. Linearization circuits have been designed to obtain linear output and to suppress zero flow offset. Linearization curves can be adjusted to match the calibration curve for a specific type of sensor, fluid, and flow range. Note that this type of flow measurement system can be applied to turbulent studies as well as full channel flow measurement. For example, Cheremisinoff[42] and Davis, Cheremisinoff, and Sambasivan[70-72] have applied "hot film" anemometry techniques to measuring wall shear stress in stratified gas-liquid and horizontal gas-liquid annular/mist flow studies.

Application to measuring total flow in a channel or pipe can follow either of two approaches:

One approach is to divide the cross-section of flow into subareas and obtain measurements by traversing the sensor or probe over the cross-section. This is the same approach that is used with a pitot tube traverse. The flow profile can be controlled so that only a single point measurement is necessary. Measurements can then be appropriately averaged to obtain an indication of the total mass flow. The total mass flow is the product of the single-point (averaged) measurement times the cross-sectional area of the pipe and an appropriate proportionality constant.

The other approach is to use an in-line transducer which maintains a flat flow profile at the sensor location. These units can be calibrated to measure total mass flow. The sensor is electrically heated as before and generates a dc voltage that is proportional to the rate of mass flow. Note that both systems described are unaffected by pressure and temperature changes.

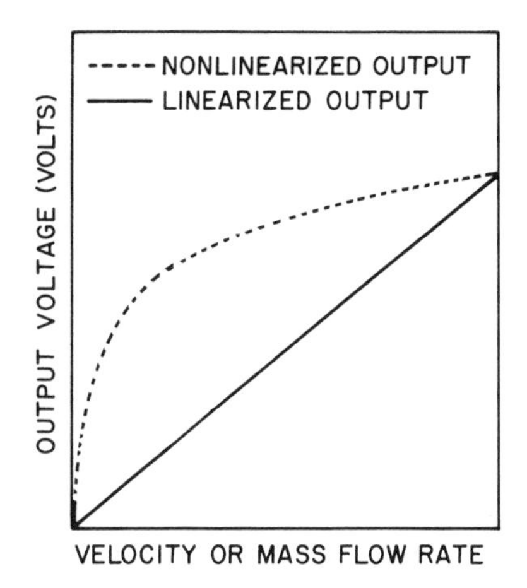

Figure 8.7 Typical nonlinear output from heated sensor exposed to fluid flow.

Limitations and Advantages

In general, mass flow transducers are intended for use in relatively clean fluids. Dirty fluids (i.e., fluids with suspended particles) or highly viscous materials can cause fouling problems and impair probe sensitivity. In liquid flow applications, probes are normally coated with a thin film (for example, sputtered quartz). This can deter excessive fouling or damage to the actual sensor and prevent electrolysis from occurring with certain types of probes.[42]

Calibration will normally change with time on miniaturized probes. Dirt or suspended particles will bake onto the sensor. Thus, new probes undergo an aging process. Calibration curves should be periodically checked to ensure no drastic changes. It is good practice to expose a new probe or transducer to actual flow conditions for several hours to a day and then check the factory calibration before use. Note that some probes are designed with a self-cleaning feature that allows burn-off contamination.

Fluid temperature variations will affect the signal output and as such must be corrected for. Corrections can be made by measuring the ambient temperature of the fluid and correcting output data via a calculation procedure. Another approach is through the use of a second sensor in the bridge circuit, so that the circuit maintains a fixed resistance ratio in the two sensors. The heated or flow sensor carries the higher current, whereas the unheated or temperature sensor carries a lower current. The unheated sensor acts like a compensating sensor which automatically corrects for errors caused by variations in the fluid's ambient or bulk temperature. If the unheated sensor is properly trimmed with balanced fixed resistors, it corrects for virtually all temperature effects.[73] In general, these systems offer the advantages of no moving parts, high sensitivity, and high accuracy in flow measurement. Air flows less than 10^{-4} SCFM can be detected with probes.[73] In standard applications, accuracy is within ±2% of reading (±0.2% or better of full scale) for the linearized signal.

Calibration

It is advisable to calibrate mass transducers in actual flow streams where possible, but this is not essential. For gas applications, these flowmeters are factory calibrated in air for standard conditions [i.e., 70° F (21.2° C) and 14.7 psi (101.3 kPa)]. Standard flowmeters can be arranged to readout flow in SCFM, SCCM, SFPM, SM/S, etc., depending on the user's requirements.

For corrosive fluids, the transducer material must be checked for compatibility. Some gases such as chlorine can greatly shorten transducer life.

NOMENCLATURE

A	Area (ft^2 or m^2)
C	Discharge coefficient
I	Current (A)
K	Constant in Eq. (8.10)
$\dot{m}$	Mass flow rate (lb/h or kg/s)
n	Constant
P	Pressure (lb_f/s $in.^2$ or N/m^2)
ΔP	Pressure drop ($lb/in.^2$ or N/m^2)
Q	Volumetric flow rate (ft^3/min or m^3/s)
q	Recycle rate (ft^3/min or m^3/s)
R	Resistance (Ω)
R_e	Cold resistance (Ω)
u	Velocity (ft/s or m/s)

Greek Letters

α	Sensor or probe parameter
β	Sensor or probe parameter
ρ	Fluid density (lb/ft^3 or g/cm^3)

9

Measurement of Open Channel and Sewer Flow

9.1 INTRODUCTION

When one surface of a flowing liquid is free of solid boundaries, the flow is referred to as "open channel." Typically, such open channel flow occurs in rivers and canals, and also in channels or pipes which are not flowing full. Open channel flow is usually measured by constructing an obstruction across the flow path and metering a parameter or characteristic variable resulting from the fluid flowing over or under the obstruction.

Open channel flow is important to many industrial applications. Effluent streams are generated by nearly every industry. Examples include breweries, slaughter houses, the pulp and paper industry, the chemical industry, and various other kinds of processing plants. The handling and cleaning of effluents is one of industry's major concerns. Accurate flow measurement has therefore become an essential operation prior to final effluent treatment.

Obtaining open channel flow measurement is not always straightforward. For example, there are numerous instances where an open sewage flow must be monitored with the only access being through a manhole. Often space limitations prevent the construction of a primary device (i.e., sluice gate or weir).

This chapter discusses various methods of metering open channel flow. Applications ranging from stream flow to measuring flows in manholes are discussed.

The use of sluice gates is not discussed in this book. A good introduction to the theory of flow measurement with sluice gates is given by Benedict,[74] and an experimental investigation on discharge coefficients for sluice gates was carried out by Henry.[75]

9.2 PRINCIPLES OF OPEN CHANNEL FLOW MEASUREMENT

Application of the energy equation to open channel flow was discussed in Chapter 2. To determine open channel flows requires establishment of a definite relationship between the level of liquid and discharge. This can be accomplished by inserting a calibrated weir or flume into the flow channel.

A weir is essentially some type of obstruction built across an open channel over which the liquid must flow. Usually there is an opening or notch on the weir plate for the fluid to flow through. There are basically three configurations of sharp-crested weirs: rectangular weirs, V-notch weirs, and Cipolletti weirs.

Rectangular Weirs

Figure 9.1 illustrates a rectangular weir with end contractions. The sheet or layer of fluid flowing over the weir is referred to as the "nappe." To determine the ideal flow rate over this configuration weir, the following assumptions are normally applied:

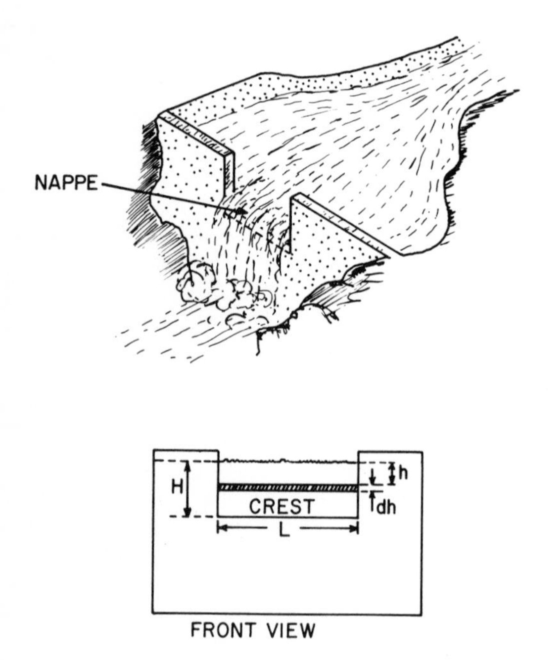

Figure 9.1 Illustrates the ideal dimensions for a fully contracted weir. The liquid flow is contracted as it passes through the rectangular opening since the ends of the weir notch are a distance from the sides of the weir pool.

1. Approach velocities are negligible.
2. The nappe is under atmospheric pressure.

The second assumption imposes the condition that the nappe be treated as a free-falling body, in which case the velocity of liquid flowing over the notch can be expressed as

$$u_h = (2gh)^{1/2} \tag{9.1}$$

where h is an incremental height of liquid (head) over the notch (refer to Figure 9.1B).

The ideal volumetric flow rate can be expressed as

$$dQ_i = u_h \, dA \tag{9.2}$$

where dA = Ldh.

Equations (9.1) and (9.2) can be combined to give

$$dQ_i = 2gh(Ldh) \tag{9.3}$$

and, integrating between the limits of 0 and H,

$$Q_i = \frac{2}{3} L(2g)^{1/2} H^{3/2} \tag{9.4}$$

Equation (9.4) states that the ideal volumetric flow rate over a rectangular weir is proportional to the 3/2 power of the liquid head.

In practice, empirical or semiempirical correlations are used to compute actual flow conditions, one of the reasons being that the first assumption used in the above derivation often has a significant effect on flow rate and cannot be neglected. The Francis formula given by Eq. (9.5) is one of the most widely used empirical expressions for flow over rectangular weirs

$$Q_a = 3.33\left(L - \frac{nH}{10}\right)\left[\left(H - \frac{u_a^2}{2g}\right)^{3/2} - \left(\frac{u_a^2}{2g}\right)^{3/2}\right] \tag{9.5}$$

where u_a is the fluid's approach velocity and n is the number of lateral contractions of the weir. Lateral contractions arise when the weir plate is not constructed over the entire width of the approach channel. For the case illustrated in Figure 9.1, n = 0 and assumption 1 becomes valid; hence, Eq. (9.5) reduces to the following:

$$Q_a = 3.33LH^{3/2} \tag{9.6}$$

Thin plastic or metal strips are normally secured along the edge of the weir notch, or the entire weir is sometimes cut from a sheet of 1/8-in. aluminum or stainless steel. Care is usually taken in selecting a weir so that at average flow conditions, the minimum head obtained is 0.2 ft. Under conditions where the overflow nappe does not clear the weir crest, accuracy in flow measurement can be significantly reduced. For relatively small flows, a V-notch weir is preferred over the rectangular design.[76]

Triangular or V-Notch Weirs

Figure 9.2 illustrates the V-notch (sometimes called triangular) weir design. By applying the same assumptions used for the flow over the rectangular weir to the geometry shown in Figure 9.2, an expression for the ideal volumetric flow rate is obtained:

$$Q_i = \frac{4L}{15H}(2g)^{1/2}H^{5/2} \tag{9.7}$$

There are a variety of empirical expressions that compensate for losses and allow reliable computations for the volumetric flow rate. The reader should refer to Refs. 77-79 for details.

Cipolletti Weirs

A Cipolletti weir is shown in Figure 9.3. This design has no major advantage over the rectangular weir. It is, however, employed in larger flow applications.

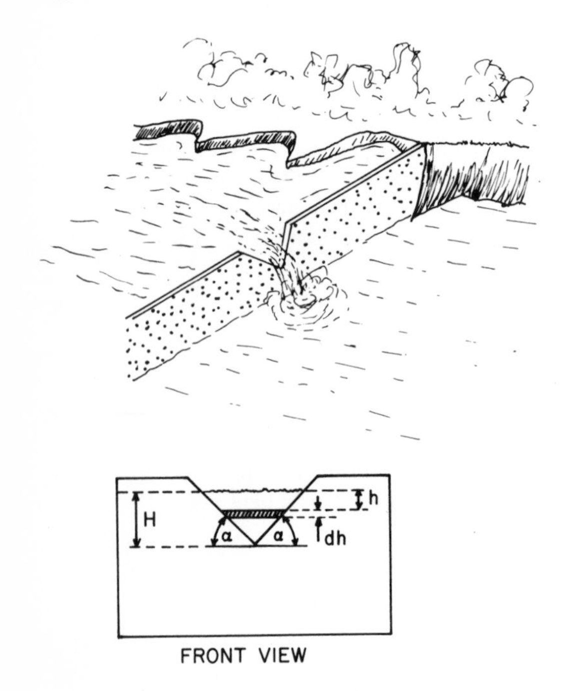

Figure 9.2 A V-notch weir.

Figure 9.3 Water flowing over a Cipolletti weir for an irrigation canal. Note that the water level behind the weir is measured by a recorder in use with a float and sheet metal stilling well. The level measured is used to compute the discharge over the weir. (Courtesy of Leupold & Stevens, Inc., Beaverton, Oreg.)

9.3 FLUMES

Flumes are specially designed open channel flow sections that provide a restriction in area which results in an increase in fluid velocity. They can be designed to be installed in a circular pipe section which has applications to open channel sewer work. Figure 9.4 is a cross-sectional drawing of a flume installed in a manhole.

Flumes have several advantages over weirs. The flow velocity through flumes is high, and so they tend to be self-cleaning systems which minimizes deposition of sediment or solids. Flumes can operate with much smaller head losses than experienced with weirs, which can be a decided advantage for irrigation and existing sewer applications.

Flumes generally require relatively smooth, near-laminar flow for proper operation. As such, they should not be installed near a sharp change in slope or near obstructions in the channel. The user must determine whether the flow is laminar at least ten pipe diameters upstream of the proposed location in sewer applications.

A Parshall flume, illustrated in Figure 9.4, is employed in applications where it is important to maintain a low head loss, or if the liquid contains a large amount of suspended solids.

9.4 DETERMINING FLOW RATE

Once the weir or flume has been installed in the flow path, the flow rate can be determined. The simplest and least sophisticated method of accomplishing this in open channels with low volume flow is by the sandbag method. This involves damming the open channel with sandbags so that the water will rise and overflow at one point. A small container is then used to collect the overflow, and a stopwatch records the collection time. From the volume of liquid collected over a recorded time interval, the volumetric flow rate can be computed. A staff gauge located on the influent side of the weir as shown in Figure 9.5 can be used to record the liquid level which can then be correlated with the measured flow rate. If sufficient data have been obtained, a correlation can be developed that allows one to compute the volumetric flow rate through the flume or weir from a measurement of the liquid level.

For large flows, the sandbag and stopwatch method is not practical. For large flows mechanical recorders are to be preferred. A float operated liquid level recorder is shown in Figure 9.6. The unit operates as follows: A float follows the liquid level as it rises and falls. The float's pulley, while turning, rotates a drum containing a chart. A clock monitor is employed to drive a pen from left to right on the instrument, producing a graphic record of the changing water levels. Depending on the gearing mechanism and type of clock drive, recording times can be varied over several hours to many days.

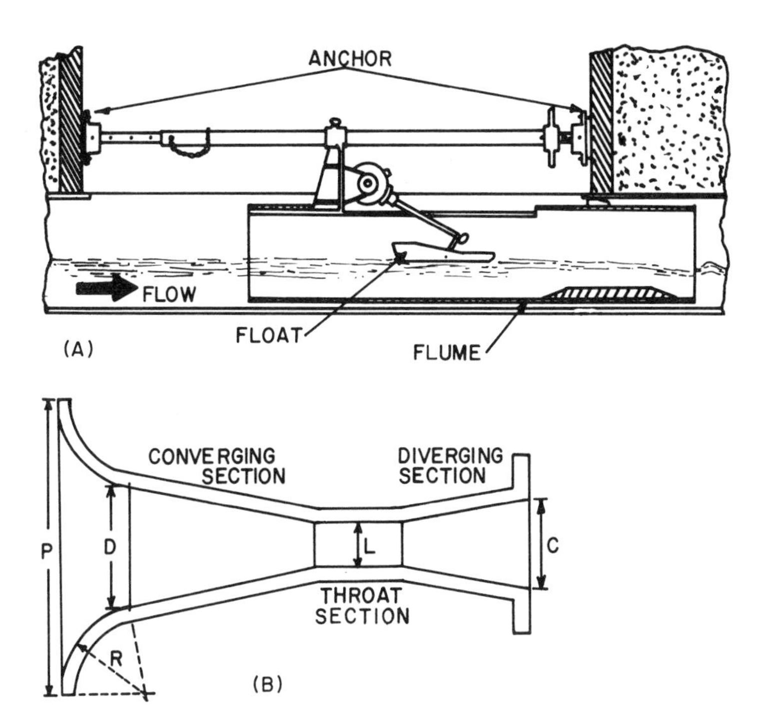

Figure 9.4 (A) Flume installed in a manhole cover. (Courtesy of NB Instruments Inc., Horsham, Pa.) (B) A Parshall flume, often employed when liquids contain a large amount of settleable solids.

The recorders are normally equipped with several scales on the same chart so that flow rate [e.g., MGD (million gallons per day)] as well as level can be read directly. Other units have been designed with built-in totalizing features.[80]

As the instrument is a float-operated mechanism, a stilling well must be used along with it. The stilling well can range from a simple metal cylinder to a concrete shaft anchored in the floor of a large instrument shelter. For flumes, stilling wells are normally located on the outside wall adjacent to the inlet area. A crossover pipe can be used to connect the bottom of the well to the side of the flume. Figure 9.7 illustrates the entire unit in operation.

It should be noted that these recorders are not universal. That is, the instrument used for recording flows is based on the particular flow characteristics of the primary measuring device and, as such, should be selected on that basis. Table 9.1 gives one manufacturer's partial listing of some of the more common devices and flow ranges experienced by their recorders.

Figure 9.5 A metal staff gauge, which resembles a ruler, can be located at the side of a flume or weir box to provide a fast visual indication of the liquid level. From a measure of the level, the flow rate can be computed.

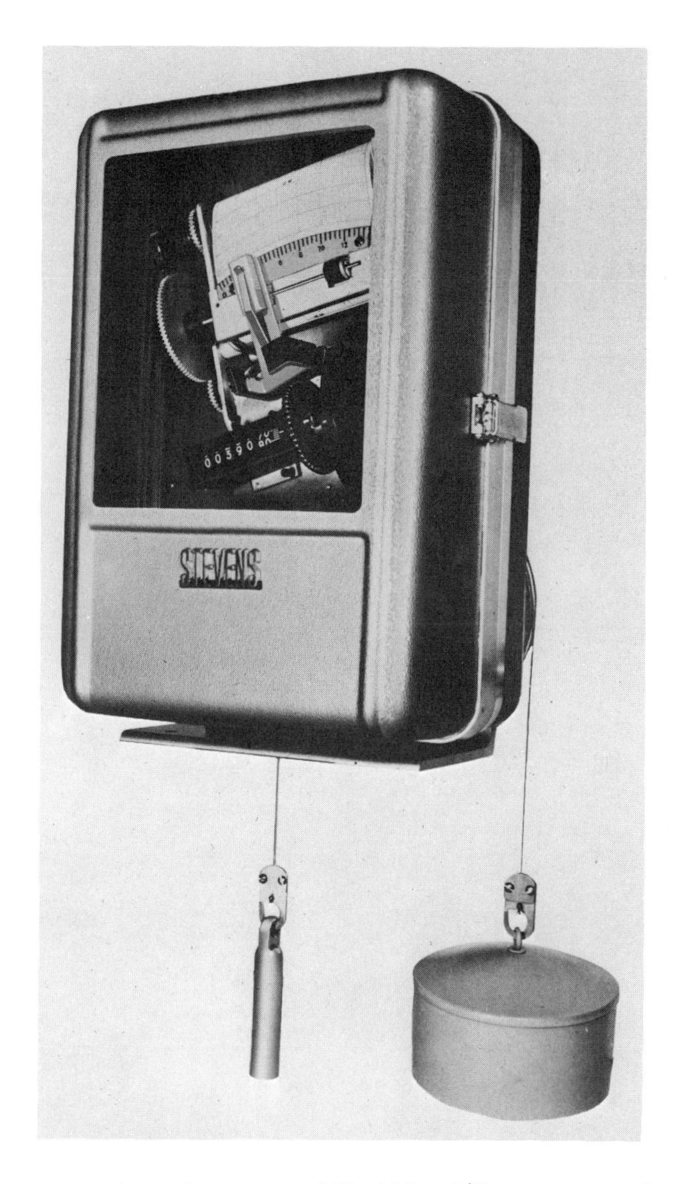

Figure 9.6 Float operated liquid level/flow rate recorder. The unit shown is used for metering open channel flow of sewage, industrial wastes, municipal water supplies, irrigation water, etc. The liquid volume is continuously totalized on a seven-digit counter, and the chart can provide up to 180 days of unattended record depending on the chart drive speed used. (Courtesy of Leupold & Stevens, Inc., Beaverton, Oreg.)

Figure 9.7 Flow recorder and stilling well in operation. Water flow over a rectangular weir is being monitored. (Courtesy of Leupold & Stevens, Inc., Beaverton, Oreg.)

Table 9.1
Partial List of One Manufacturer's Total Flowmeters—Measuring Devices and Flow Ranges[a]

Weir or flume	Max. flow (MGD)	Totalizer multiplier (to determine) vol. in gallons)	Sampling rate of optional actuating switch (gallons through weir of flume between samples)			Recommended float diameter (in.)
			10-lobe cam	20-lobe cam	25-lobe cam	
22 1/2° V-notch weir	0.350	100	1000	500	400	7
	0.140	40	400	200	160	9
	0.070	20	200	100	80	10
	0.035	10	100	50	40	12
	0.0175	5	50	25	20	14
	0.014	4	40	20	16	14
45° V-notch weir	1.750	500	5000	2500	2000	6
	0.700	200	2000	1000	800	8
	0.350	100	1000	500	400	9
	0.175	50	500	250	200	10
	0.140	40	400	200	160	10
	0.070	20	200	100	80	12
60° V-notch weir	1.750	500	5000	2500	2000	7
	1.400	400	4000	2000	1600	7
	0.700	200	2000	1000	800	8
	0.350	100	1000	500	400	9
	0.175	50	500	250	200	10
	0.140	40	400	200	160	12
	0.070	20	200	100	80	14

Table 9.1
(continued)

Weir or flume	Max. flow (MGD)	Totalizer multiplier (to determine vol. in gallons)	Sampling rate of optional actuating switch (gallons through weir of flume between samples)			Recommended float diameter (in.)
			10-lobe cam	20-lobe cam	25-lobe cam	
90° V-notch weir	7.000	2000	20,000	10,000	8000	6
	3.500	1000	10,000	5000	4000	7
	1.750	500	5000	2500	2000	8
	1.400	400	4000	2000	1600	8
	0.700	200	2000	1000	800	9
	0.350	100	1000	500	400	10
	0.175	50	500	250	200	12
120° V-notch weir	35.000	10,000	100,000	50,000	40,000	6
	17.500	5000	50,000	25,000	20,000	6
	7.000	2000	20,000	10,000	8000	6
	3.500	1000	10,000	5000	4000	7
12-in. Cipolletti weir	0.700	200	2000	1000	800	12
	0.350	100	1000	500	400	14
2-ft Cipolletti weir	1.750	500	5000	2500	2000	10
12-in. rectangular weir without end contraction	3.500	1000	10,000	5000	4000	6
	1.750	500	5000	2500	2000	8
	1.400	400	4000	2000	1600	9

2-ft rectangular weir	3.500	1000	10,000	5000	4000	8
without end contraction	1.400	400	4000	2000	1600	12
12-in. rectangular weir	0.700	200	2000	1000	800	12
with end contraction	0.350	100	1000	500	400	14
2-ft rectangular weir	3.500	1000	10,000	5000	4000	8
with end contraction	1.750	500	5000	2500	2000	10
	1.400	400	4000	2000	1600	12
1-in. Parshall flume	0.140	40	400	200	160	9
	0.070	20	200	100	80	12
	0.035	10	100	50	40	14
2-in. Parshall flume	0.350	100	1000	500	400	9
	0.175	50	500	250	200	10
	0.070	20	200	100	80	14
3-in. Parshall flume	0.700	200	2000	1000	800	8
	0.350	100	1000	500	400	9
	0.175	50	500	250	200	12
	0.140	40	400	200	160	14
6-in. Parshall flume	3.500	1000	10,000	5000	4000	6
	1.750	500	5000	2500	2000	7
	1.400	400	4000	2000	1600	8
	0.700	200	2000	1000	800	9
	0.350	100	1000	500	400	12
9-in. Parshall flume	7.000	2000	20,000	10,000	8000	6
	3.500	1000	10,000	5000	4000	6
	1.750	500	5000	2500	2000	8
	1.400	400	4000	2000	1600	8
	0.700	200	2000	1000	800	10
	0.350	100	1000	500	400	14

Table 9.1
(continued)

Weir or flume	Max. flow (MGD)	Totalizer multiplier (to determine vol. in gallons)	Sampling rate of optional actuating switch (gallons through weir of flume between samples)			Recommended float diameter (in.)
			10-lobe cam	20-lobe cam	25-lobe cam	
12-in. Parshall flume	7.000	2000	20,000	10,000	8000	6
	3.500	1000	10,000	5000	4000	7
	1.750	500	5000	2500	2000	8
	0.700	200	2000	2000	800	12
18-in. Parshall flume	7.000	2000	20,000	10,000	8000	6
	3.500	1000	10,000	5000	4000	8
	1.750	500	5000	2500	2000	10
	0.700	200	2000	1000	800	14
2-ft Parshall flume	17.500	5000	50,000	25,000	20,000	6
	14.000	4000	40,000	20,000	16,000	6
	7.000	2000	20,000	10,000	8000	7
	3.500	1000	10,000	5000	4000	9
	1.400	400	4000	2000	1600	12
3-ft Parshall flume	35.000	10,000	100,000	50,000	40,000	6
	14.000	4000	40,000	20,000	16,000	6
5-ft Parshall flume	70.000	20,000	200,000	100,000	80,000	6
	35.000	10,000	100,000	50,000	40,000	6

6-in. Palmer-Bowlus flume	0.350	100	1000	500	400	12
	0.175	50	500	250	200	14
	0.140	40	400	200	160	14
8-in. Palmer-Bowlus flume	0.700	200	2000	1000	800	10
	0.350	100	1000	500	400	12
	0.175	50	500	250	200	14
10-in. Palmer-Bowlus flume	1.400	400	4000	2000	1600	9
	0.700	200	2000	1000	800	10
	0.350	100	1000	500	400	12
12-in. Palmer-Bowlus flume	1.750	500	5000	2500	2000	9
	1.400	400	4000	2000	1600	9
	0.700	200	2000	1000	800	10
	0.350	100	1000	500	400	12
15-in. Palmer-Bowlus flume	3.500	1000	10,000	5000	4000	6
	1.750	500	5000	2500	2000	8
	1.400	400	4000	2000	1600	9
6-in. Leopold-Lagco flume	0.140	40	400	200	160	14
8-in. Leopold-Lagco flume	0.350	100	1000	500	400	12
	0.175	50	500	250	200	14
10-in. Leopold-Lagco flume	0.700	200	2000	1000	800	9
	0.350	100	1000	500	400	12
12-in. Leopold-Lagco flume	1.400	400	4000	2000	1600	8
	0.700	200	2000	1000	800	10

Table 9.1
(continued)

Weir or flume	Max. flow (MGD)	Totalizer multiplier (to determine vol. in gallons)	Sampling rate of optional actuating switch (gallons through weir of flume between samples)			Recommended float diameter (in.)
			10-lobe cam	20-lobe cam	25-lobe cam	
15-in. Leopold-Lagco flume	1.750	500	5000	2500	2000	8
	1.400	400	4000	2000	1600	8
	0.700	200	2000	1000	800	10
	0.350	100	1000	500	400	14
18-in. Leopold-Lagco flume	3.500	1000	10,000	5000	4000	7
	1.750	500	5000	2500	2000	8
	1.400	400	4000	2000	1600	9
	0.700	200	2000	1000	800	12
	0.350	100	1000	500	400	14

[a]Courtesy of Leupold & Stevens, Inc., Beaverton, Oreg.

9.5 FLOW MEASUREMENT IN MANHOLES

Manhole flow measuring installations are usually temporary systems. For this reason, instrumentation is portable and preferably operated independent of an external power source. Weirs and flumes are used in sewer flow metering applications. As indicated earlier, flow through the primary measuring device is a function of the liquid level and, as such, the discharge at any given instant can be determined from a single level measurement. The relationship between level and fluid discharge is termed "rating." Rating curves (refer to Figure 9.8) and tabulated rating values for standard weirs and flumes are available from manufacturers.

Flumes are generally selected for use with untreated sewage as they tend to be self-cleaning systems and operate at a lower head loss than weirs do. Figure 9.4 illustrates the use of a flume in this type of application.

Weirs are often considered the least expensive and simplest devices to install in a manhole for flow measurement. One type of design is illustrated in Figure 9.9. The arrangement shown is useful for situations where portable flumes face difficult installation problems; for example, in manholes with a junction or a manhole where the incoming or outgoing pipes, or both, are not in line with the channel.[81] The unit shown consists of an arrangement of sliding plastic bars that can be positioned to dam or block the open channel under the bulkhead. Slotted mounting holes and an integral bubble level allow adjustment of the weir to match flow conditions.

In the upper corners of the bulkhead are pointed threaded rods which, when tightened against the manhole walls, support the bulkhead until a head

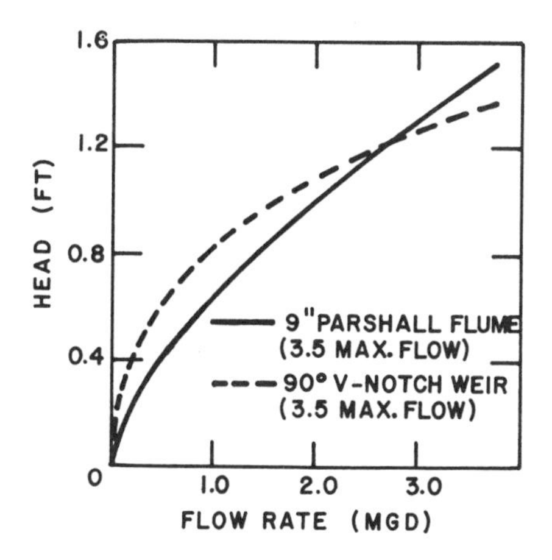

Figure 9.8 Typical rating curves. Shown are the rating curves for a 90° V-notch weir and a 9-in. (23-cm) Parshall flume. (Courtesy of Leupold & Stevens, Inc., Beaverton, Oreg.)

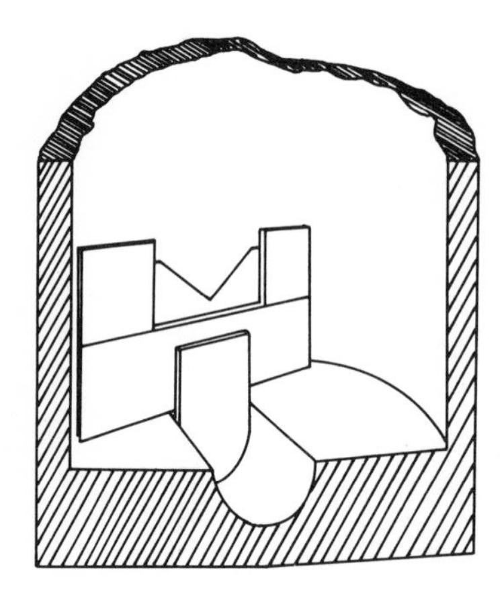

Figure 9.9 A V-notch, bulkhead weir for measuring flows in manholes. (Courtesy of NB Instruments, Inc., Horsham, Pa.)

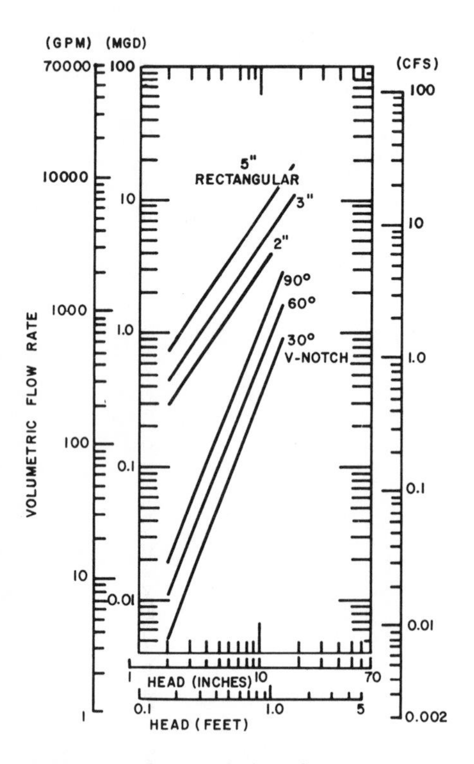

Figure 9.10 Graph for obtaining size and type of weir. (Courtesy of The Foxboro Co., Foxboro, Mass.)

of fluid builds up behind the weir. A watertight seal between the bulkhead and manhole walls can be achieved with a temporary sealing compound.

9.6 SIZING THE PRIMARY DEVICE

If maximum and minimum flow conditions are known, the primary unit can be sized based on empirical or theoretical expressions using a safety factor. For weirs the Francis expression [Eq. (9.5)] along with the recommended dimensions in Figure 9.1 can be used to size the unit.

The Foxboro Co.[82] has prepared graphs that can be used in the selection of size and type of weir. Figure 9.10 can be used for rectangular and V-notch weirs. The range of head, H, and construction details can be obtained from Figure 9.1.

NOMENCLATURE

A	Area (ft^2 or m^2)
g	Acceleration of gravity (32.2 ft/s^2 or 981.4 cm/s^2)
H	Height from bottom of weir notch to liquid surface, head (ft or m)
h	Head (ft or m)
L	Width or length (ft or m)
Q	Volumetric flow rate (ft^3/s or m^3/s)
u_a	Approach velocity (ft/s or m/s)
u_h	Velocity over weir (ft/s or m/s)

Subscripts

a	Refers to actual conditions
i	Refers to ideal or theoretical conditions

References

1. Park, R. A., "Dispersion Resin Handbook: Outline of Processing Principles and Raw Material Characteristics Relating to Dispersion Resin Systems," Firestone Plastics Co.
2. Skelland, A. H. P., Non-Newtonian Flow and Heat Transfer, Wiley, New York, 1967.
3. Bird, R. B., W. E. Stewart, and E. N. Lightfoot, Transport Phenomena, Wiley, New York, 1960.
4. Uyehara, O. A., and K. M. Watson, Nat. Petrol. News, Tech. Sec., 36, R764 (Oct. 4, 1944).
5. Foust, A. S., L. A. Wenzel, C. W. Clump, L. Maus, and L. B. Anderson, Principles of Unit Operations, Wiley, New York, 1960.
6. Hedstrom, B. O. A., "Flow of Plastic Materials in Pipes," Ind. Eng. Chem., 44, 651 (1952).
7. Merrill, E. W., A coaxial cylinder viscometer for the study of fluids under high velocity gradients, J. Colloid Sci., 9, 7 (1954).
8. Merrill, E. W., Pseudoplastic flow: Viscometry, correlations of shear stress vs. shear rate, and predictions of laminar flow in tubes, J. Colloid Sci., 11, 1 (1956).
9. Streeter, V. L. (Ed.), Handbook of Fluid Dynamics, McGraw-Hill, New York, 1961.
10. McAdams, W. H., Heat Transmission, 3rd ed., McGraw-Hill, New York, 1954.
11. Knudsen, J. G., and D. L. Katz, Fluid Dynamics and Heat Transfer, McGraw-Hill, New York, 1958.
12. Prandtl, L., Z. Ver. Deut. Ing., 77, 105 (1933).
13. Nikuradse, J., Forsch. Ver. Deut. Ing., p. 356 (1932).
14. Von Karman, T., N.A.C.A., T.M., 611 (1931).
15. Deissler, R. E., "Analytical and Experimental Investigation of Adiabatic Turbulent Flow in Smooth Tubes," NACA-TN-2138 (1952).

16. Moody, L. F., Friction factors for pipe flow, Trans. A.S.M.E., 66 (1944).
17. Colebrook, C. F., and D. White, J. Inst. Civil Engrs. (London), 11, 133 (1938-1939).
18. King, H. W., and E. F. Brater, Handbook of Hydraulics, McGraw-Hill, New York, 1963.
19. Jeppson, R. W., Analysis of Flow in Pipe Networks, Ann Arbor Science Publns., Ann Arbor, Mich., 1976.
20. Vennard, J. K., Elementary Fluid Mechanics, Wiley, New York, 1961.
21. Metcalf & Eddy, Inc., Wastewater Engineering, McGraw-Hill, New York, 1972.
22. Metcalf, L., and H. P. Eddy, American Sewage Practice, 2nd ed., Vol. 1, McGraw-Hill, New York, 1928.
23. Dodge, D. W., and A. B. Metzner, Amer. Inst. Chem. Eng. J., 5, 189-204 (1959); 8, 143 (1962).
24. Metzner, A. B., in Handbook of Fluid Dynamics (V. L. Streeter, Ed.), McGraw-Hill, New York, 1961.
25. Bogue, D. C., and A. B. Metzner, Ind. Eng. Chem. (Fundamentals), 2(2), 143-149 (1963).
26. Clapp, R. M., "International Developments in Heat Transfer," Pt. 3, 652-661, D-159, D-211-5, ASME, New York (1961).
27. Pai, S. I., Viscous Flow Theory, Vol. 2: Turbulent Flow, Van Nostrand Reinhold, New York, 1957.
28. Brodkey, R. S., J. Lee, and R. C. Chase, Amer. Inst. Chem. Eng. J., 7, 392-393 (1961).
29. Bingham, E. C., Fluidity and Plasticity, McGraw-Hill, New York, 1922.
30. Herschel, W. H., and R. Bulkley, Proc. Amer. Soc. Testing Materials, 26, 621 (1926).
31. Herschel, W. H., and R. Bulkley, Kolloid-Z., 39, 291 (1926).
32. Crowley, P. R., and A. S. Kitzes, Ind. Eng. Chem., 49, 888-892 (1957).
33. Buckingham, E., Proc. Amer. Soc. Testing Materials, 21, 1154-1161 (1921).
34. Venturi, G. B., "Recherches experimentales sur le principe de la communication latérale du mouvement dans les fluides appliqué à l'explication de différents phénomènes hydrauliques" (1797).
35. Herchel, C., The Venturi water meter, Trans. Amer. Soc. Civil Engrs. (1887).
36. Aeroquip Corp., Jackson, Mich., "Barco-Venturi Flow Measurement System" (1975).
37. Fluid Meters, Their Theory and Application, 5th ed., ASME, New York, 1959.
38. Flowmeter Computation Handbook, ASME, New York, 1961.
39. Cheremisinoff, N. P., and R. Niles, Survey of fluid flow measurement techniques and fundamentals, Water & Sewage Works, 122, No. 12 (1975).

40. Fribance, A. E., Industrial Instrumentation Fundamentals, McGraw-Hill, New York, 1962.
41. Spink, L. K., Principles and Practice of Flow Meter Engineering, 8th ed., The Foxboro Co., Foxboro, Mass., 1958.
42. Cheremisinoff, N. P., "An Experimental and Theoretical Investigation of Horizontal Stratified and Annular Two-Phase Flow with Heat Transfer," Ph.D. Thesis, Dept. Chem. Eng., Clarkson College of Technology, Potsdam, N.Y. (1977).
43. Tuve, G. L., Mechanical Engineering Experimentation, McGraw-Hill, New York, 1961.
44. W. A. Kates Co., Deerfield, Ill., "Automatic Flow Rate Controllers," Brochure 701 (1978).
45. Froude, W., Discharge of elastic fluids under pressure, Proc. Inst. Civil Engrs. (London), 6 (1847).
46. Bean, H. S., and S. R. Beitler, Some results from research on flow nozzles, Trans. A.S.M.E. (April 1938).
47. Bean, H. S., S. R. Beitler, and R. E. Sprenkle, Discharge coefficients of long radius flow nozzles when used with pipe wall pressure taps, Trans. A.S.M.E. (July 1941).
48. Folsom, R. G., Nozzle coefficients for free and submerged discharge, Trans. A.S.M.E. (April 1939).
49. Kallen, H. P. (Ed.), Handbook of Instrumentation and Controls, McGraw-Hill, New York, 1961.
50. Arnberg, B. T., Review of critical flowmeters for gas flow measurements, Trans. A.S.M.E., 84D (1962).
51. Holman, J. P., Experimental Methods for Engineers, McGraw-Hill, New York, 1971.
52. Hersey Products Inc., Spartanburg, S.C., "Liquid Meters," Catalog 67710M (1978).
53. RCM Industries, Orinda, Calif., "Product Specification Sheet—Series 7000 Flow Gage," Sheet S-700-677 (1978).
54. C-E Invalco, Combustion Engineering Inc., Tulsa, Okla., "Hoverflo Bearingless Turbine Flowmeter,"Bruchure IVC-251-A4 (1978).
55. C-E Invalco, Combustion Engineering Inc., Tulsa, Okla., "Sanitary Turbine Flowmeters," Brochure IVC-255-A1 (1978).
56. Flow Technology, Inc., Phoenix, Ariz., "Product Data from Flow Technology Inc.: Turbine Flowmeters for Oil Field Automation," Bulletin PB-753 (1978).
57. Flow Technology, Inc., Phoenix, Ariz., "Turbine Flow Transducers," Bulletin TP-771 (1978).
58. Flow Technology, Inc., Phoenix, Ariz., "Retractable Turbo-Probes," Bulletin TR-761 (1978).
59. C-E Invalco, Combustion Engineering Co., Tulsa, Okla., "Flowmeters: PT-Series, RT-Series," Bulletin IVC-252 (1978).

60. Foxboro Corp., Foxboro, Mass., "Magnetic Flow Measurement Systems," Bulletin E10-D (July 1975).
61. Foxboro Corp., Foxboro, Mass., "Magnetic Flow Meter Application Information," Technical Information T127-71e (May 1968).
62. Foxboro Corp., Foxboro, Mass., "Magnetic Flow Transmitter Materials Guide," Technical Information T127-71f (July 1973).
63. Bestobell Mobrey Limited, Slough, Bucks., England, "Doppler Flowmeter: Ultrasonic Flow Measurement of Liquids or Slurries in Pipes," Bulletin MF1 (May 1977).
64. Hersey Products Inc., Spartanburg, S.C., "Doppler Flow Measurement Systems," Bulletin DFT-1 (Mar. 1978).
65. Cochrane Environmental Systems, Pa., Specifications sheet on "Uni Pac Monitor—Model M-1" (1978).
66. Cochrane Environmental Systems, Pa., Specifications sheet on "Noncontacting Characterized Flow Transmitter," Model M-3 (1978).
67. Flo-Tron, Inc., Paterson, N.J., "Model 10 Low Flow Linear Mass Flowmeter," Bulletin F-10 (1978).
68. Masnik, W., "Flo-Tron Microflow Meter," paper presented at International Symposium on Automotive Technology and Automation, Wolfsburg, W. Germany (Sept. 12-15, 1977).
69. Flo-Tron, Inc., Paterson, N.J., "Model 10M-Microflow Meter," Bulletin F-15 (1978).
70. Davis, E. J., N. P. Cheremisinoff, and G. Sambasivan, "Heat and Momentum Transfer Analogies for Horizontal Stratified Two-Phase Flow," State of Art Report to A.I.Ch.E.—Design Institute for Multiphase Processing, Clarkson College of Technology, Potsdam, N.Y. (June 1976).
71. Davis, E. J., and N. P. Cheremisinoff, "Heat Transfer to Two-Phase Stratified Gas-Liquid Flow," Report 5 to A.I.Ch.E.—Design Institute for Multiphase Processing, Clarkson College of Technology, Potsdam, N.Y. (June 1977).
72. Davis, E. J., and N. P. Cheremisinoff, "Stratified Two-Phase Flow: Pressure Drop and Holdup," Report 4 to the A.I.Ch.E.—Design Institute for Multiphase Processing, Clarkson College of Technology, Potsdam, N.Y. (June 1977).
73. Datametrics Inc., Wilmington, Mass., "Gould Datametrics, Series-1000, Mass Flow Transducers," Bulletin 1050B (May 1977).
74. Benedict, R. P., Fundamentals of Temperature, Pressure, and Flow Measurements, Wiley, New York, 1969.
75. Henry, H. R., "A Study of Flow from a Submerged Sluice Gate," M.S. Thesis, State University of Iowa, Ames, Iowa (Feb. 1950).
76. Leupold & Stevens, Inc., Beaverton, Oreg., "Measuring Open Channel Wastewater Flows," Stevens Application Notes, No. 3 (1977).
77. Bakhmeteff, B. A., Hydraulics of Open Channels, McGraw-Hill, New York, 1932.

78. Chow, V. T., Open Channel Hydraulics, McGraw-Hill, New York, 1959.
79. Rouse, H., Fluid Mechanics for Hydraulic Engineers, Dover, New York, 1961.
80. Leupold & Stevens, Inc., Beaverton, Oregon, "One Recorder—Many Uses," Stevens Application Notes, No. 2 (1978).
81. NB Instruments, Inc., Horsham, Pa., "Bulkhead Weir for Manholes," Bulletin 708R (1976).
82. The Foxboro Co., Foxboro, Mass., "Open Channel Flow Measurement with Parshall Flumes and Weirs," Technical Information Sheet T19-12a (Feb. 1974).

appendix A

Unit Conversions

To convert from:	To units of:	Multiply by:
	GENERAL CONVERSIONS	
acres	square feet	43,560
	square meters	4047
	square miles	1.562×10^{-3}
	square yards	4840
angstroms (Å)	inches	3.937×10^{-9}
	centimeters	9.999×10^{-9}
centigrams	grams	0.01
centiliters	liters	0.01
centimeters	feet	0.032808
	inches	0.3938
	meters	0.01
	mils	393.7
	millimeters	10
circular mils	square centimeters	5.067×10^{-6}
	square inches	7.854×10^{-7}
	square mils	0.7854

To convert from:	To units of:	Multiply by:
cubic centimeters	cubic feet	3.531×10^{-5}
	cubic inches	6.102×10^{-2}
	cubic meters	10^{-6}
	cubic yards	1.308×10^{-6}
	gallons	2.642×10^{-4}
	liters	10^{-3}
	pint (liq.)	2.113×10^{-3}
	quarts (liq.)	1.057×10^{-3}
	ounces (U.S. fluid)	0.033814
cubic feet	cubic centimeters	2.832×10^{-4}
	cubic inches	1728
	cubic meters	0.02832
	cubic yards	0.03704
	gallons	7.481
	liters	28.32
	pints (liq.)	59.84
	quarts (liq.)	29.92
ft^3/min	cm^3/s	472.0
	gal/s	0.1247
	liters/s	0.4720
	lb of water/min	62.4
	gal/min	448.83
	million gal/day	0.64632
cubic inches	cubic centimeters	16.39
	cubic feet	5.787×10^{-4}
	cubic meters	1.639×10^{-5}
	cubic yards	2.143×10^{-5}
	gallons	4.329×10^{-3}
	liters	1.639×10^{-2}

To convert from:	To units of:	Multiply by:
cubic inches (cont.)	pints (liq.)	0.03463
	quarts (liq.)	0.01732
cubic meters	cubic centimeters	10^6
	cubic feet	35.31
	cubic inches	61,023
	cubic yards	1.308
	gallons	264.2
	liters	10^3
	pints (liq.)	2113
	quarts (liq.)	1057
cubic yards	cubic centimeters	7.646×10^5
	cubic feet	27
	cubic inches	46,656
	cubic meters	0.7646
	gallons	202.0
	liters	764.6
	pints (liq.)	1616
	quarts (liq.)	807.9
days	minutes	1440
	seconds	86,400
gallons (U.S.)	ounces (U.S. fluid)	128
	cubic centimeters	3785
	cubic feet	0.1337
	cubic inches	231
	cubic meters	3.785×10^{-3}
	cubic yards	4.951×10^{-3}
	liters	3.785
	pints (liq.)	8
	quarts (liq.)	4

To convert from:	To units of:	Multiply by:
gal/min	ft^3/s	2.228×10^{-3}
	liters/s	0.06308
grams	kilograms	10^{-3}
	milligrams	10^3
	ounces	0.03527
	ounces (Troy)	0.03215
	poundals	0.07093
	pounds	2.205×10^{-3}
g/cm^3	lb/ft^3	62.43
	$lb/in.^3$	0.03613
	lb/mil·ft	3.405×10^{-7}
	lb/gal	8.34
g/liter	g/cm^3	9.99973×10^{-4}
	parts/million	1000
	lb/ft^3	0.06243
hours (h)	minutes	60
	seconds	3600
inches	centimeters	2.540
	mils	10^3
kg/m^3	g/cm^3	10^{-3}
	lb/ft^3	0.06243
	$lb/in.^3$	3.613×10^{-5}
	lb/mil·ft	3.405×10^{-10}
kilometers	centimeters	10^5
	feet	3281
	meters	10^3
	miles	0.6214
	yards	1093.6
km/h	cm/s	27.78

To convert from:	To units of:	Multiply by:
km/h (cont.)	ft/min	54.68
	ft/s	0.9113
	knots/h	0.5396
	meters/min	16.67
	miles/h	0.6214
$km \cdot h^{-1} \cdot s^{-1}$	$cm \cdot s^{-2}$	27.78
	$ft \cdot s^{-2}$	0.9113
	$m \cdot s^{-2}$	0.2778
	$miles \cdot h^{-1} \cdot s^{-1}$	0.6214
liters	cubic centimeters	10^{3}
	cubic feet	0.03531
	cubic inches	61.02
	cubic meters	10^{-3}
	cubic yards	1.308×10^{-3}
	gallons	0.2642
	pints (liq.)	2.113
	quarts (liq.)	1.057
liters/min	ft^{3}/s	5.885×10^{-4}
	gal/s	4.403×10^{-3}
meters	feet	3.2808
	inches	39.37
	kilometers	10^{-3}
	millimeters	10^{3}
	yards	1.0936
	angstroms	10^{10}
	miles	6.2137×10^{4}
micrograms (μg)	grams	10^{-6}
microliters (μl)	liters	10^{-6}
micrometers (μm)	meters	10^{-6}

To convert from:	To units of:	Multiply by:
miles	centimeters	1.609×10^{5}
	feet	5280
	kilomters	1.6093
	yards	1760
miles/h	cm/s	44.70
	ft/m	88
	ft/s	1.467
	km/h	1.6093
	m/min	26.82
milligrams	grams	10^{-3}
milliliters	liters	10^{-3}
millimeters	centimeters	0.1
	inches	0.03937
	mils	39.37
mils	centimeters	0.002540
	inches	10^{-3}
	microns	25.40
months	days	30.42
	hours	730
	minutes	43,800
	seconds	2.628×10^{6}
square centimeters	circular mils	1.973×10^{5}
	square feet	1.076×10^{-3}
	square inches	0.1550
	square meters	10^{-6}
	square millimeters	100
square feet	acres	2.296×10^{-5}
	square centimeters	929.0
	square inches	144

To convert from:	To units of:	Multiply by:
square feet (cont.)	square meters	0.09290
	square miles	3.587×10^{-8}
	square yards	1/9
square inches	circular mils	1.273×10^{6}
	square centimeters	6.452
	square feet	6.944×10^{-3}
	square mils	10^{6}
	square millimeters	645.2
	square yards	7.71605×10^{-4}
square kilometers	acres	247.1
	square feet	10.76×10^{6}
	square meters	10^{6}
	square miles	0.3861
	square yards	1.196×10^{6}
square meters	acres	2.471×10^{-4}
	square feet	10.764
	square miles	3.861×10^{-7}
	square yards	1.196
square miles	acres	640
	square feet	27.88×10^{6}
	square kilometers	2.590
	square yards	3.098×10^{6}
square millimeters	circular mils	1.973×10^{3}
	square centimeters	0.01
	square inches	1.550×10^{-3}
square mils	circular mils	1.273
	square centimeters	6.452×10^{-6}
	square inches	10^{-6}
square yards	acres	2.066×10^{-4}

To convert from:	To units of:	Multiply by:
square yards (cont.)	square feet	9
	square meters	0.8361
	square miles	3.228×10^{-7}
tons (long)	kilograms	1016
	pounds	2240
tons (metric)	kilograms	10^3
	pounds	2205
tons (short)	kilograms	907.2
	pounds	2000
tons (short)/ft^2	kg/m^2	9765
	$lb/in.^2$	13.89
yards	centimeters	91.44
	feet	3
	inches	36
	meters	0.9144
	UNITS OF FORCE	
dynes	$g \cdot cm/s^2$	1
	newtons	10^{-5}
	poundals	7.2330×10^{-5}
	pound-force	2.248×10^{-6}
	grams	1.020×10^{-3}
pounds-force (lb_f)	dynes	4.4482×10^5
	newtons	4.4482
	poundals	32.174
newtons (N)	$kg \cdot m/s^2$	1
	dynes	10^5
	poundals	7.2330
	pounds-force	2.2481×10^{-1}

To convert from:	To units of:	Multiply by:
poundals	$lb_m \cdot ft/s^2$	1
	dynes	1.3826×10^4
	newtons	1.3826×10^{-1}
	pounds-force	3.1081×10^{-2}
	grams	14.10
	UNITS OF PRESSURE, MOMENTUM, FLUX	
atmospheres (atm)	centimeters of mercury	76.0
	inches of mercury	29.921
	feet of water	33.90
	kg/m^2	10.333×10^3
	$lb_f/in.^2$	14.696
	$tons/ft^2$	1.058
	dyn/cm^2	1.0133×10^6
	N/m^2	1.0133×10^5
	$poundals/ft^2$	6.8087×10^4
	lb_f/ft^2	2.1162×10^3
	millimeters of mercury	760
$dyne/cm^2$	$g \cdot cm^{-1} \cdot s^{-2}$	1
	N/m^2	10^{-1}
	$poundals/ft^2$	6.7197×10^{-2}
	lb_f/ft^2	2.0886×10^{-3}
	$lb_f/in.^2$	1.450×10^{-5}
	atmospheres	9.8692×10^{-7}
	millimeters of mercury	7.5006×10^{-4}
	inches of mercury	2.9530×10^{-5}
	bars	1
N/m^2	$kg \cdot m^{-1} \cdot s^{-2}$	1
	$dyne/cm^2$	10

To convert from:	To units of:	Multiply by:
N/m^2 (cont.)	$poundals/ft^2$	6.7197×10^{-1}
	lb_f/ft^2	2.0886×10^{-2}
	$lb_f/in.^2$	1.4504×10^{-4}
	atmospheres	9.8692×10^{-6}
	millimeters of mercury	7.5006×10^{-3}
	inches of mercury	2.9530×10^{-4}
inches of mercury	atmospheres	0.03342
	feet of water	1.133
	kg/cm^2	0.0345
	kg/m^2	345.3
	millimeters of mercury	25.40
	lb_f/ft^2	70.73
	$lb_f/in.^2$	0.4912
inches of water	atmospheres	0.002458
	inches of mercury	0.07355
	kg/m^2	25.40
	$oz/in.^2$	0.5781
	lb_f/ft^2	5.204
	$lb_f/in.^2$	0.03613
lb_f/ft^2	feet of water	0.01602
	kg/m^2	4.882
	$lb_f/in.^2$	6.944×10^{-3}
$lb_f/in.^2$	atmospheres	0.06804
	feet of water	2.307
	inches of mercury	2.036
	kg/cm^2	0.0703
	kg/m^2	703.1
	lb_f/ft^2	144

To convert from:	To units of:	Multiply by:
$lb_f/in.^2$ (cont.)	g/cm^2	70.307
	millimeters of mercury	51.715
	UNITS OF WORK, ENERGY, TORQUE	
British thermal units (Btu)	kg·cal	0.2520
	$ft \cdot lb_f$	777.5
	hp·h	3.927×10^{-4}
	joules	1054
	kg·m	107.5
	kW·h	2.928×10^{-4}
	ergs	1.0550×10^{10}
	ft·poundals	2.5036×10^{4}
	calories	2.5216×10^{2}
calories (gram)	Btu	3.9685×10^{-3}
	$ft^3 \cdot atm$	0.001469
	$ft \cdot lb_f$	3.0874
	W·h	0.0011628
ergs	dyn·cm	1
	Btu	9.486×10^{-11}
	$ft \cdot lb_f$	7.376×10^{-8}
	g·cm	1.020×10^{-3}
	joules	10^{-7}
	kg·cal	2.390×10^{-11}
	kg·m	1.020×10^{-8}
	hp·h	3.7251×10^{-14}
	kW·h	2.7778×10^{-14}
	ft·poundals	2.3730×10^{-6}
ft·poundals	$lb_m \cdot ft^2/s^2$	1
	ergs	4.2140×10^{5}

To convert from:	To units of:	Multiply by:
ft·poundals (cont.)	joules	4.2140×10^{-2}
	ft·lb_f	3.1081×10^{-2}
	calories	1.0072×10^{-2}
	Btu	3.9942×10^{-5}
	hp·h	1.5698×10^{-8}
	kW·h	1.1706×10^{-8}
horsepower-hours (hp·h)	Btu	2547
	ft·lb_f	1.980×10^{6}
	joules	2.684×10^{6}
	kg·cal	641.7
	kg·m	2.737×10^{5}
	kW·h	0.74570
	ergs	2.6845×10^{13}
	ft·poundals	6.3705×10^{7}
	calories	6.4162×10^{5}
	UNITS OF POWER	
Btu/min	ft·lb_f/s	12.96
	horsepower	0.02356
	kilowatts	0.01757
	watts	17.57
horsepower (hp)	Btu/min	42.44
	ft·lb_f/min	33,000
	ft·lb_f/s	550
	horsepower (metric)	1.014
	kg·cal/min	10.70
	kilowatts	0.7457
	watts	745.7
kilowatts (kW)	Btu/min	56.92

To convert from:	To units of:	Multiply by:
kilowatts (kW) (cont.)	$ft \cdot lb_f/min$	4.425×10^4
	$ft \cdot lb_f/s$	737.6
	horsepower	1.341
	kg·cal/min	14.34
	watts	10^3
$ft \cdot lb_f/min$	Btu/min	1.286×10^{-3}
	$ft \cdot lb_f/s$	0.01667
	horsepower	3.030×10^{-5}
	kg·cal/min	3.241×10^{-4}
	kilowatts	2.260×10^{-5}
$ft \cdot lb_f/s$	Btu/min	7.717×10^{-2}
	horsepower	1.818×10^{-3}
	kg·cal/min	1.945×10^{-2}
	kilowatts	1.356×10^{-3}
	Btu/h	4.6275
	watts	1.35582
watts (W)	Btu/min	0.05692
	ergs/s	10^7
	$ft \cdot lb_f/min$	44.26
	$ft \cdot lb_f/s$	0.7376
	horsepower	1.341×10^{-3}
	kg·cal/min	0.01434
	kilowatts	10^{-3}
	UNITS FOR THERMAL CONDUCTIVITY	
$Btu \cdot h^{-1} \cdot ft^{-1} \cdot {}^\circ F^{-1}$	$erg \cdot s^{-1} \cdot cm^{-1} \cdot K^{-1}$	1.7307×10^5
	$W \cdot m^{-1} \cdot K^{-1}$	1.7307
	$lb_m \cdot ft \cdot s^{-3} \cdot {}^\circ F^{-1}$	6.9546
	$lb_f \cdot s^{-1} \cdot {}^\circ F^{-1}$	2.1616×10^{-1}

To convert from:	To units of:	Multiply by:
$Btu \cdot h^{-1} \cdot ft^{-1} \cdot {}^\circ F^{-1}$	$cal \cdot s^{-1} \cdot cm^{-1} \cdot K^{-1}$	4.1365×10^{-3}
$cal \cdot s^{-1} \cdot cm^{-1} \cdot K^{-1}$	$erg \cdot s^{-1} \cdot cm^{-1} \cdot K^{-1}$	4.1840×10^{7}
	$W \cdot m^{-1} \cdot K^{-1}$	4.1840×10^{2}
	$lb_m \cdot ft \cdot s^{-3} \cdot {}^\circ F^{-1}$	1.6813×10^{3}
	$lb_f \cdot s^{-1} \cdot {}^\circ F^{-1}$	5.2256×10^{1}
	$Btu \cdot h^{-1} \cdot ft^{-1} \cdot {}^\circ F^{-1}$	2.4175×10^{2}
$erg \cdot s^{-1} \cdot cm^{-1} \cdot K^{-1}$	$g \cdot cm \cdot s^{-3} \cdot K^{-1}$	1
	$W \cdot m^{-1} \cdot K^{-1}$	10^{-5}
	$lb_m \cdot ft \cdot s^{-3} \cdot {}^\circ F^{-1}$	4.0183×10^{-5}
	$lb_f \cdot s^{-1} \, {}^\circ F^{-1}$	1.2489×10^{-6}
	$cal \cdot s^{-1} \cdot cm^{-1} \cdot K^{-1}$	2.3901×10^{-8}
	$Btu \cdot h^{-1} \cdot ft^{-1} \cdot {}^\circ F^{-1}$	5.7780×10^{-6}
$lb_m \cdot ft \cdot s^{-3} \cdot {}^\circ F^{-1}$	$erg \cdot s^{-1} \cdot cm^{-1} \cdot K^{-1}$	2.4886×10^{4}
	$W \cdot m^{-1} \cdot K^{-1}$	2.4886×10^{-1}
	$lb_f \cdot s^{-1} \cdot {}^\circ F^{-1}$	3.1081×10^{-2}
	$cal \cdot s^{-1} \cdot cm^{-1} \cdot K^{-1}$	5.9479×10^{-4}
	$Btu \cdot h^{-1} \cdot ft^{-1} \cdot {}^\circ F^{-1}$	1.4379×10^{-1}
$lb_f \cdot s^{-1} \cdot {}^\circ F^{-1}$	$erg \cdot s^{-1} \cdot cm^{-1} \cdot K^{-1}$	8.0068×10^{5}
	$W \cdot m^{-1} \cdot K^{-1}$	8.0068
	$lb_m \cdot ft \cdot s^{-3} \cdot {}^\circ F^{-1}$	32.174
	$cal \cdot s^{-1} \cdot cm^{-1} \cdot K^{-1}$	1.9137×10^{-2}
	$Btu \cdot h^{-1} \cdot ft^{-1} \cdot {}^\circ F^{-1}$	4.6263
$W \cdot m^{-1} \cdot K^{-1}$	$kg \cdot m \cdot s^{-3} \cdot K^{-1}$	1
	$erg \cdot s^{-1} \cdot cm^{-1} \cdot K^{-1}$	10^{5}
	$lb_m \cdot ft \cdot s^{-3} \cdot {}^\circ F^{-1}$	4.0183
	$lb_f \cdot s^{-1} \cdot {}^\circ F^{-1}$	1.2489×10^{-1}
	$cal \cdot s^{-1} \cdot cm^{-1} \cdot K^{-1}$	2.3901×10^{-3}
	$Btu \cdot h^{-1} \cdot ft^{-1} \cdot {}^\circ F^{-1}$	5.7780×10^{-1}

To convert from:	To units of:	Multiply by:
UNITS FOR HEAT TRANSFER COEFFICIENT		
$Btu\cdot h^{-1}\cdot ft^{-2}\cdot °F^{-1}$	$cal\cdot cm^{-2}\cdot s^{-1}\cdot K^{-1}$	1.3571×10^{-4}
	$g\cdot cm^{-3}\cdot K^{-1}$	5.6782×10^{3}
	$W\cdot cm^{-2}\cdot K^{-1}$	5.6782×10^{-4}
	$W\cdot m^{-2}\cdot K^{-1}$	5.6782
	$lb_m\cdot s^{-3}\cdot °F^{-1}$	6.9546
	$lb_f\cdot ft^{-1}\cdot s^{-1}\cdot °F^{-1}$	2.1616×10^{-1}
$cal\cdot cm^{-2}\cdot s^{-1}\ K^{-1}$	$g\cdot s^{-3}\cdot K^{-1}$	4.1840×10^{7}
	$W\cdot m^{-2}\cdot K^{-1}$	4.1840×10^{4}
	$lb_m\cdot s^{-3}\cdot °F^{-1}$	5.1245×10^{4}
	$lb_f\cdot ft^{-1}\cdot s^{-1}\cdot °F^{-1}$	1.5928×10^{3}
	$W\cdot cm^{-2}\cdot K^{-1}$	4.1840
	$Btu\cdot ft^{-2}\cdot h^{-1}\cdot °F^{-1}$	7.3686×10^{3}
$g\cdot s^{-3}\cdot K^{-1}$	$Btu\cdot ft^{-2}\cdot h^{-1}\cdot °F^{-1}$	1.7611×10^{-4}
	$cal\cdot cm^{-2}\cdot s^{-1}\cdot K^{-1}$	2.3901×10^{-8}
	$lb_m\cdot s^{-3}\cdot °F^{-1}$	1.2248×10^{-3}
	$lb_f\cdot ft^{-1}\cdot s^{-1}\cdot °F^{-1}$	3.8068×10^{-5}
	$W\cdot cm^{-2}\cdot K^{-1}$	10^{-7}
	$W\cdot m^{-2}\cdot K^{-1}$	10^{-3}
$kg\cdot s^{-3}\cdot K^{-1}$	$Btu\cdot ft^{-2}\cdot h^{-1}\cdot °F^{-1}$	1.7611×10^{-1}
	$cal\cdot cm^{-2}\cdot s^{-1}\cdot K^{-1}$	2.3901×10^{-5}
	$g\cdot s^{-3}\cdot K^{-1}$	10^{3}
	$lb_m\cdot s^{-3}\cdot °F^{-1}$	1.2248
	$lb_f\cdot ft^{-1}\cdot s^{-1}\cdot °F^{-1}$	3.8068×10^{-2}
	$W\cdot m^{-2}\cdot K^{-1}$	1
	$W\cdot cm^{-2}\cdot K^{-1}$	10^{-4}
$lb_m\cdot s^{-3}\cdot °F^{-1}$	$Btu\cdot ft^{-2}\cdot h^{-1}\cdot °F^{-1}$	1.4379×10^{-1}
	$cal\cdot cm^{-2}\cdot s^{-1}\cdot K^{-1}$	1.9514×10^{-5}

To convert from:	To units of:	Multiply by:
$lb_m \cdot s^{-3} \cdot {}^\circ F^{-1}$ (cont.)	$g \cdot s^{-3} \cdot K^{-1}$	8.1647×10^2
	$lb_f \cdot ft^{-1} \cdot s^{-1} \cdot {}^\circ F^{-1}$	3.1081×10^{-2}
	$W \cdot m^{-2} \cdot K^{-1}$	8.1647×10^{-1}
	$W \cdot cm^{-2} \cdot K^{-1}$	8.1647×10^{-5}
$lb_f \cdot ft^{-1} \cdot s^{-1} \cdot {}^\circ F^{-1}$	$Btu \cdot ft^{-2} \cdot h^{-1} \cdot {}^\circ F^{-1}$	4.6263
	$cal \cdot cm^{-2} \cdot s^{-1} \cdot K^{-1}$	6.2784×10^{-4}
	$g \cdot s^{-3} \cdot K^{-1}$	2.6269×10^4
	$lb_m \cdot s^{-3} \cdot {}^\circ F^{-1}$	32.1740
	$W \cdot m^{-2} \cdot K^{-1}$	2.6269×10^1
	$W \cdot cm^{-2} \cdot K^{-1}$	2.6269×10^{-3}
$W \cdot cm^{-2} \cdot K^{-1}$	$Btu \cdot ft^{-2} \cdot h^{-1} \cdot {}^\circ F^{-1}$	1.7611×10^3
	$cal \cdot cm^{-2} \cdot s^{-1} \cdot K^{-1}$	2.3901×10^{-1}
	$g \cdot s^{-3} \cdot K^{-1}$	10^7
	$lb_m \cdot s^{-3} \cdot {}^\circ F^{-1}$	1.2248×10^4
	$lb_f \cdot ft^{-1} \cdot s^{-1} \cdot {}^\circ F^{-1}$	3.8068×10^2
	$W \cdot m^{-2} \cdot K^{-1}$	10^4

UNITS FOR VISCOSITY, DENSITY TIMES DIFFUSIVITY

To convert from:	To units of:	Multiply by:
centipoise (cP)	poise	10^{-2}
	$kg \cdot m^{-1} \cdot s^{-1}$	10^{-3}
	$lb_m \cdot ft^{-1} \cdot s^{-1}$	6.7197×10^{-4}
	$lb_f \cdot s/ft^2$	2.0886×10^{-5}
	$lb_m \cdot ft^{-1} \cdot h^{-1}$	2.4191
$kg \cdot m^{-1} \cdot s^{-1}$	poise	10
	$lb_m \cdot ft^{-1} \cdot s^{-1}$	6.7197×10^{-1}
	$lb_f \cdot s/ft^2$	2.0886×10^{-2}
	centipoise	10^3
	$lb_m \cdot ft^{-1} \cdot h^{-1}$	2.4191×10^3

To convert from:	To units:	Multiply by:
$lb_m \cdot ft^{-1} \cdot s^{-1}$	poises	1.4882×10^{1}
	$kg \cdot m^{-1} \cdot s^{-1}$	1.4882
	$lb_f \cdot s/ft^2$	3.1081×10^{-2}
	centipoises	1.4882×10^{3}
	$lb_m \cdot ft^{-1} \cdot h^{-1}$	3600
$lb_f \cdot s/ft^2$	poise	4.7880×10^{2}
	$kg \cdot m^{-1} \cdot s^{-1}$	4.7880×10^{1}
	$lb_m \cdot ft^{-1} \cdot s^{-1}$	32.1740
	centipoise	4.7880×10^{4}
	$lb_m \cdot ft^{-1} \cdot h^{-1}$	1.1583×10^{5}
$lb_m \cdot ft^{-1} \cdot h^{-1}$	centipoise	4.1338×10^{-1}
	$lb_f \cdot s/ft^2$	8.6336×10^{-6}
	$lb_m \cdot ft^{-1} \cdot s^{-1}$	2.7778×10^{-4}
	$kg \cdot m^{-1} \cdot s^{-1}$	4.1338×10^{-4}
	poises	4.1338×10^{-3}
poise (P)	$g \cdot cm^{-1} \cdot s^{-1}$	1
	$kg \cdot m^{-1} \cdot s^{-1}$	0.10
	$lb_m \cdot ft^{-1} \cdot s^{-1}$	6.7197×10^{-2}
	$lb_f \cdot s/ft^2$	2.0886×10^{-3}
	centipoise	10^{2}
	$lb_m \cdot ft^{-1} \cdot h^{-1}$	2.4191×10^{2}

appendix B

Solutions to Study Problems

ANSWERS TO QUESTIONS IN CHAPTER 1

1.1 Viscosity directly proportional to shearing force (see Section 1.2).

1.2 $\tau_{xz} = \frac{\mu}{g_c} \frac{du}{dx}$

1.3 Time-independent fluids, time-dependent fluids, and viscoelastic fluids.

1.4 Refers to materials that exhibit a yield value.

1.5 See Eq. (1.6); the viscosity of a non-Newtonian fluid assumed to follow Newton's law of viscosity.

1.6 Power law expression, Eq. (1.7).

1.7 Fluids that exhibit an increase in apparent viscosity with increasing shear rate.

1.8 Refers to an isothermal gel-sol-gel transformation or breakdown of a reversible colloidal gel (see Section 1.3, discussion of time-dependent fluids).

1.9 $0 \leq n < 1$

1.10 $n > 1$

1.11 The concavity of the hysteresis loop differs.

1.12 Curve of τ versus shear rate or viscosity versus shear rate that describes fluid properties as a function of time.

1.13 Flour dough, petroleum jellies, nylons.

1.14 Ketchup, toothpaste.

1.15 Black liquors from paper industry, ketchup, mayonnaise.

1.16 Shear stress or force that must be applied to induce flow.

1.17 Rheological and volumetric dilatancy.

1.18 Zero velocity at the fluid-solid boundary.

1.19 Rheopexy is a form of thixotropy; dilatant fluids increase in viscosity with increasing shear.

1.20 Viscosity decreases with increasing temperature.

1.21 Viscosity increases with increasing temperature.

1.22 Viscosity increases with increasing pressure.

1.23 Viscosity increases with increasing pressure.

1.24 Density, average velocity of molecules, and mean free path of molecules.

1.25 The science of deformation and flow of matter.

SOLUTIONS TO PROBLEMS IN CHAPTER 2

2.1 (a) From a mass balance

$$\Sigma Q_i = Q_1 + Q_2 - Q_3 - Q_4 = 0$$

$$0.2 + Q_2 - 0.42 - 0.07 = 0$$

$$Q_2 = 0.29 \text{ m}^3/\text{s}$$

(b) $\bar{u}_i = Q_i/A_i$, $A = \frac{1}{4}\pi(D_1)^2$

$\bar{u}_1 = 10.96$ m/s, $\bar{u}_2 = 11.7$ m/s, $\bar{u}_3 = 10.23$ m/s, $\bar{u}_4 = 15.35$ m/s

2.2 (a) $Q = \bar{u}/A$

$$= (15 \text{ m/s})/[\tfrac{1}{4}\ (0.1905)^2]$$

$$= 526.3 \text{ m}^3/\text{s}$$

(b) $W = \rho Q = (1 \text{ g/cc})(526.3 \text{ m}^3/\text{s})(10^6 \text{ cm}^3/\text{m}^3)/3600$

$$= 146.2 \text{ kg/hr}$$

(c) From Eq. (2.1),

$$\bar{u}_1 = \bar{u}_2 \frac{D_2}{D_1}$$

$$= 15\left(\frac{7.5}{3}\right) = 37.5 \text{ m/s}$$

(d) $Re = \dfrac{Du}{\mu}$

$$= \frac{(0.0762\ m)(37.5\ m/s)(1\ g/cm^3)}{(1.09\ g/cm\cdot s)}(10^4\ cm^2/m^2)$$

$= 26,215$

Since Re > 2100, the flow is turbulent.

2.3 (a) $H_1 = (h_L)_{entrance} + z_B + P_B/\rho g + \dfrac{u_B^2}{2g} + h_{L_{AB}}$

(b) $H_1 = (h_L)_{entrance} + z_C + P_C/\rho g + \dfrac{u_C^2}{2g} + h_{L_{AB}} + h_T$

(c) $H_2 = z_D + \dfrac{P_D}{\rho g} + \dfrac{u_D^2}{2g} - h_{L_{DE}} + h_P$

2.4 (a) From Eq. (2.37),

$h_L/L = f\bar{u}^2/2gD$

$u = 0.02/[\frac{1}{4}\pi(0.127)^2] = 1.58\ m/s$

$$Re = \frac{Du\rho}{\mu} = \frac{(0.127)(1.58)(1)}{(1.09)}(10^4) = 1839$$

Since Re < 2100, the flow is laminar; use Eq. (2.41)

$f = 64/Re = 64/1839 = 0.0348$

$$\frac{h_L}{L} = (0.0348)\frac{(1.58)^2}{(2)(9.814)(0.127)} = 3.481 \times 10^{-2}\ ft\ H_2O/ft\ pipe$$

or

$$\frac{h_L}{L} = \left(3.481 \times 10^{-2}\ \frac{ft\ H_2O}{ft}\right)(0.4335) = 1.509 \times 10^{-2}\ psi/ft$$

$$= 341.3\ N/m^2/m$$

(b) $\Delta P = g\rho h_L$

$$\Delta P = \left(32.2\ \frac{ft}{s^2}\right)\left(62.4\ \frac{lb_m}{ft^3}\right)\left(3.481 \times 10^{-2}\ \frac{ft}{ft}\right)(70\ m)\left(\frac{ft}{0.3048\ m}\right)$$

$= 1.6063 \times 10^4\ poundals/ft^2$

$= 3.47\ psi$

2.5 From Eq. (2.14),

$$\tau_w = \frac{R}{2}\frac{\Delta P}{L} \qquad L = 70 \text{ m} = 229.6 \text{ m}$$

$$\tau_w = \frac{(2.5/12)}{2}\frac{(3.47)}{229.6} = 1.574 \times 10^{-3} \text{ psi}$$

$$= 10.85 \text{ N/m}^2$$

2.6 $A = \frac{1}{4}\pi(1/12)^2 = 5.45 \times 10^{-3} \text{ ft}^2$

$$\bar{u} = \frac{5200}{(3600)(67)(5.45 \times 10^{-3})} = 3.96 \text{ ft/s}$$

From Eq. (2.60),

$$f = 16/\text{Re}'$$

where

$$\text{Re}' = \frac{D^n \bar{u}^{2-n} \rho}{k}(8)\left(\frac{n}{6n + 2}\right)^n$$

$$= \frac{\left(\frac{1}{12}\right)^{0.75}(3.96)^{1.25}(67)}{22}(8)\left(\frac{0.75}{4.5}\right)^{0.75}$$

$$= 5.51$$

Since the flow is clearly laminar, Eq. (2.57) can be used to compute the frictional pressure drop:

$$\Delta P = 2(22)\left(\frac{3 \times 0.75 + 1}{0.75}\right)^{0.75}\frac{(3.96)^{0.75}(10)}{(32.174)\left(\frac{0.5}{12}\right)^{1.75}}$$

$$= 3.001 \times 10^4 \text{ lb}_f/\text{ft}^2$$

$$= 208 \text{ psi}$$

2.7 (a) Use Eq. (2.59); to compute the centerline velocity, r = 0 and Eq. (2.59) reduces to:

$$u = \left(\frac{3n + 1}{n + 1}\right)\bar{u}$$

$$= \frac{3(0.75) + 1}{0.75 + 1}\bar{u} = 1.86\bar{u}$$

or

$$u_{r=0} = 1.86(3.96) = 7.35 \text{ ft/s}$$

(b) From solution (a), $u_{r=0} = 1.86\,\bar{u}$. See the accompanying tabulation.

Tube diameter (in.)	u at r=0 (ft/s)	$\bar{u}$ (ft/s)
0.5	29.41	15.81
1.0	7.35	3.96
2.0	1.84	0.99
3.0	0.82	0.44

2.8 (a) $H_T = 18 = f\frac{L}{D}\frac{\bar{u}^2}{2g} + K_L\frac{\bar{u}^2}{2g}$

From Table 2.3, $K_L = 5.6$. Then

$$18 = \frac{\bar{u}^2}{2g}\left(\frac{300}{2/12}\,f + 5.6\right)$$

$$1159.2 = \bar{u}^2\,(1800\,f + 5.6) \qquad *$$

A trial and error solution can be performed. The Moody diagram (Figure 2.5) can be used by choosing a value for the Reynolds number, obtaining the friction factor from the plot, and computing a value of $\bar{u}$ from Re. With values for $\bar{u}$ and f from a guess of Re, the right-hand side (RHS) of the above equation (*) is solved. If it matches the LHS (left-hand side), the guess is correct; if not, a new value of Re is chosen and the calculation repeated until convergence within a specified accuracy is obtained. The procedure is illustrated below:

Trial 1. Choose $Re = 10^5$; from Moody plot, $f = 0.0173$. Then

$$\bar{u} = Re\,\mu/D\rho$$

$$= 10^5\left(7.324 \times 10^{-4}\,\frac{lb_m}{ft\cdot s}\right)\Big/\left(\frac{2}{12}\,ft\right)\left(62.4\,\frac{ft}{s}\right)$$

$$= 7.04\ ft/s$$

$$LHS = (7.04)^2\,[1800(0.0173) + 5.6]$$

$$= 1822.4$$

Since RHS ≠ LHS, a new value for Re is chosen. Calculations are tabulated as shown here.

Trial no.	Re	$\bar{u}$	f	RHS of eq. (*)
1	1.0×10^5	7.04	0.0173	1822.4
2	6.0×10^4	4.23	0.0193	720.4
3	9.0×10^4	6.34	0.0178	1512.9
4	7.8×10^4	5.49	0.0181	1152.0

The fourth trial agrees within 0.6%; hence,

$\bar{u} = 5.49$ ft/s and

$$Q = \bar{u}A = (5.49)\left[\frac{1}{4}\ (2/12)^2\right] = 0.1198\ \text{ft}^3/\text{s}$$
$$= 53.8\ \text{GPM}$$

(b) h_L from valve $= K_L \dfrac{\bar{u}^2}{2g}$

$$= 5.6\,\frac{(5.49)^2}{2(32.2)}$$
$$= 2.62$$

Hence, contribution to the head loss from the valve equals

$$\frac{2.62}{18} \times 100 = 14.6\%$$

SOLUTIONS TO PROBLEMS IN CHAPTER 3

3.1 (a) $A_1 = \frac{1}{4}\pi(3/12)^2 = 4.909 \times 10^{-2}\ \text{ft}^2$

$$\bar{u} = (140\ \text{GPM})(2.228 \times 10^{-3})/4.909 \times 10^{-2} = 6.35\ \text{ft/s}$$

$$\text{Re} = \frac{D\bar{u}\rho}{\mu} = \frac{(3/12)(6.35)(62.4)}{2.36}(3600) = 151{,}213$$

(b) Use Eq. (3.5) to solve for MA_2 (assume $C \simeq 0.99$):

$$Q_a = 140\ \text{GPM}\ (2.228 \times 10^{-3}) = 0.312\ \text{ft}^3/\text{s}$$

$$0.312 = 0.99\ MA_2 \sqrt{\frac{2(32.174)}{62.4}(9.0)(144)}$$

$$MA_2 = 8.621 \times 10^{-3}$$

Solve for the throat diameter (d):

$$MA_2 = \frac{A_2}{\sqrt{1 - (A_2/A_1)^2}} = 8.62 \times 10^{-3}$$

$$A_2 = \frac{1}{4}\pi d^2 = 8.757 \times 10^{-3} \quad \text{or}$$

$$d = 0.1056 \text{ ft}$$

$$= 1.267 \text{ in. } (3.22 \text{ cm})$$

Using the formulas given in Figure 3.3, the other critical dimensions are computed to be:

$$b = d = 1.267 \text{ in.}$$

$$C = d/2 = 0.633 \text{ in.}$$

$$r_2 = 3.5d = 4.434 \text{ in.}$$

$$r_1 = 1.375D = 4.125 \text{ in.}$$

(c) From Eqs. (3.22) and (3.23)

$$Re_T = \frac{\rho u_m d}{\mu} = \frac{\dot{m} d}{(\pi d^2/4)\mu} = \frac{4\dot{m}}{\pi d \mu}$$

$$\dot{m} = Q_a \rho = (0.312)(62.4) = 19.47 \text{ lb/s}$$

$$Re_T = \frac{4(19.47)}{\pi(0.1056)(2.36)}(3600) = 358{,}076$$

Note: Since $Re_T > 2 \times 10^5$, our value for C was reasonable.

3.2 Since $\Delta P = 72 \text{ in. } H_2O \times 5.204 = 374.7 \text{ lb}_f/\text{ft}^2$

$$\beta = d/D = 7/12 = 0.583$$

Either from interpolation of values from Table 3.2 or Eq. (3.6), obtain a value for M:

$$M = [1 - (A_2/A_1)^2]^{-1/2} = (1 - \beta^4)^{-1/2}$$

$$= [1 - (0.583)^2]^{-1/2} = 1.2308$$

Assume C = 0.98. Then, using Eq. (3.5):

$$Q_a = (0.98)(1.2308)\left[\frac{1}{4}\pi(7/12)^2\right]\sqrt{\frac{2(32.174)}{0.8(62.4)}(374.7)}$$

$$= 7.084 \text{ ft}^3/\text{s} \quad \text{or} \quad 53 \text{ GPH}$$

$$\dot{m} = Q_a \rho_a = (7.084)(62.4 \times 0.8) = 353.6 \text{ lb/s}$$

3.3 Since $\Delta P = 150$ in. $H_2O \times 5.204 = 780.6\ lb_f/ft^2$

$$= 5.421 \text{ psi}$$

$$P_2 = 352.8 + 14.7 - 5.421 = 362.1 \text{ psi}$$

$$\frac{P_2}{P_1} = \frac{362.1}{367.5} = 0.985$$

Compute the expansion factor from Eq. (3.11):

$\beta = 6/10 = 0.6$

$$Y = \left[(0.985)^{2/1.47}\left(\frac{1.47}{0.47}\right)\frac{1 - (0.985)^{0.47/1.47}}{1 - 0.985} \quad \frac{1 - (0.6)^4}{1 - (0.6)^4(0.985)^{2/1.47}}\right]$$

$$= 0.9817$$

Using Eq. (3.10) with C = 0.98 (note, from Table 3.2, M = 1.0719),

$$A_2 = \frac{1}{4}\pi(0.5)^2 = 0.1963 \text{ ft}^2$$

$$= (0.9817)(0.98)(1.0719)(0.1963)\sqrt{2(32.174)(0.081)(780.6)}$$

$$\dot{m}_a = 12.91\ lb_m/s$$

3.4 Since $\Delta P = 180 \times 0.03613 = 6.503$ psi

$$\frac{P_1 - P_2}{P_1} = \frac{6.503}{204.7} = 3.177 \times 10^{-2}$$

Use Eq. (3.20) to compute the expansion factor:

$$Y = 1 - [0.333 + 1.145(0.7^2 + 0.7^6 + 12(0.7)^{13})] \frac{3.177 \times 10^{-2}}{(32.2)(.081).55}$$

$$= 0.957$$

Assume C = 0.62; from Table 3.2, M = 1.1472. Use Eq. (3.21) to compute the mass flow rate:

$$k = CM = 0.62(1.1472) = 0.7113$$

$$d = \beta D = 5.6 \text{ in.}$$

$$A_2 = \frac{1}{4}\pi(5.6/12)^2 = 0.1710 \text{ ft}^2$$

$$\dot{m}_a = (0.957)(0.7113)(0.1710)\sqrt{2(32.174)(0.55 \times 0.081)(6.503)(144)}$$

$$\dot{m}_a = 6.032 \text{ lb/s} \times 3600 = 21,717 \text{ lb/hr}$$

3.5 Use Eq. (3.18) to compute the expansion factor:

$$Y = 1 - [0.41 + 0.35(0.7)^4] \frac{3.177 \times 10^{-2}}{(32.2)(.081)(0.55)}$$

$$= 0.989$$

From Eq. (3.21),

$$\dot{m}_a = (0.989)(0.7113)(0.1710)\sqrt{2(32.174)(0.55 \times 0.081)(6.503)(144)}$$

$$= 22{,}431 \text{ lb/hr}$$

3.6 (a) Use Eq. (3.24),

$$\left(\frac{P_2}{P_1}\right)_{crit} = \left(\frac{2}{2.53}\right)^{1.53/0.53}$$

$$= 0.507$$

(b) Use Eq. (3.25) to solve for A_2 and then d (for air, R = 53.35 $ft \cdot lb_f/lb_m \cdot {}^\circ R$):

$$4.5 = A_2(735)(144)\sqrt{\frac{(2)(32.174)}{(53.35)(630)}\left[\frac{1.53}{2.53}\left(\frac{2}{2.53}\right)^{2/0.53}\right]^{1/2}}$$

$$A_2 = 1.947 \times 10^{-3} \text{ ft}^2$$

$$d = \sqrt{\frac{4}{\pi}(1.947 \times 10^{-3})} \times 12$$

$$= 0.597 \text{ in.}$$

3.7 The assumption used to solve Problem 3.6 is that the static pressure is approximately equal to the stagnation pressure. This can be proved as follows: From the ideal gas law, mass flow upstream is

$$\dot{m} = \left(\frac{P_1}{RT_1}\right)_{ST} A_1 u_1$$

Subscript ST refers to static conditions. In terms of the stagnation temperature,

$$u_1 = \sqrt{2g_c C_P(T_{10} - T_{1_{ST}})}$$

Subscript 0 refers to stagnation conditions. Combining the two equations,

$$\dot{m} = \left(\frac{P_1}{RT_1}\right)_{ST} A_1 \sqrt{2g_c C_P (T_{10} - T_{1_{ST}})}$$

$$C_P = 0.241 \ \frac{\text{Btu}}{\text{lb}\cdot{}^\circ\text{F}} \times 778.16 = 187.54 \ \frac{\text{ft}\cdot\text{lb}_f}{\text{lb}_m\cdot{}^\circ\text{F}}$$

Solving for T_{10},

$$4.5 = \frac{(735)(144)}{(53.35)(630)} \ \frac{1}{4}\pi\left(\frac{4}{12}\right)^2 \sqrt{2(32.174)(187.54)(630 - T_{1_{ST}})}$$

$$T_{10} = 629.98^\circ\text{R} \quad \text{or} \quad \Delta T = 0.0222\,^\circ\text{F}$$

Since the temperature difference is so small, the upstream velocity must be small. This can be shown from the second equation:

$$u_1 = \sqrt{2(32.174)(187.54)(0.0222)}$$
$$= 16.4 \text{ ft/s}$$

The pressure difference corresponding to this velocity is 1.4 psia, which is negligible compared to 300 psia. Hence, the assumption for this diameter pipe is valid and calculations for Problem 3.6 justified.

Index